Collecting
Farm
ANTIQUES
Identification and Values

by Lar Hothem

BOOKS AMERICANA INC.

ISBN 0-89689-035-X

Cover photo courtesy of

Rebecca Lowers
(private collection)

Photographer-William Folwell, ca. 1900
Parkersburg, West Virginia

Accoutrements to early American living: Half-log and heavy peg-leg seating; sheepskin cover; spinning wheel in right background; many glazed stoneware containers.

Photo courtesy THE NATIONAL COLONIAL FARM of the Accokeek Foundation, Accokeek, Maryland; Kathleen Carlson, photographer.

Farm out-buildings. Center, the summer-kitchen or out-kitchen with brick-constructed fireplace and bake-oven at lower right of structure and chimney. Center left, the "necessary" or out-house. Far left, the smokehouse, where colonists smoked beef and pork.

Photos courtesy THE NATIONAL COLONIAL FARM of the Accokeek Foundation, Accokeek, Maryland; Kathleen Carlson, photographer.

Man in Colonial garb works at a shaving horse, using a drawknife to thin a board. Note how foot pressure tightens shaving head on piece being worked. Frame-constructed building rises in background.

Acknowledgements

The writer wishes to express deep thanks and appreciation to the following individuals and/or institutions for considerable and invaluable assistance in putting this book together. If any contributor has been overlooked, the unintentional error will be corrected in subsequent editions.

My thanks to: Michael R. Regan, Editor, and Edward A. Babka, Publisher, *The Antiques Journal* (P.O. Box 1046, Dubuque, Iowa 52001) for generous permission to use some material previously published in the pages of AJ in article format, and also for the agreement to allow reproduction of certain photographs used in the same context.

Also, gratitude to Tom Porter, President of GARTH'S AUCTIONS, Inc., Delaware, Ohio, for permission to use auction listings and results in many of the 165 separate farm-item subsections of this book. These accurate listings add a range and depth that could not be found elsewhere. Thanks also for objects listed in various sales, this due to Mike Clum, Auctioneer (Thornville, Ohio) for examples in several quality farm-item auctions.

Appreciation to L. Edward Gastellum, Superintendent, and Elizabeth Bauer, Curator, Hubbell Trading Post National Historic Site, Ganado, Arizona, for permission to use photographs obtained on those premises. Thanks also to them for researching the purpose and manufacture-marks of several pieces of farm machinery on display at Hubbell Trading Post.

Thanks to Edwin Wingfield, "The Windmill Man", of Hamilton, Illinois, for contributing copy for the section on windmill weights. Special thanks to Ivan Glick, Public Relations, SPERRY/NEW HOLLAND, New Holland, Pennsylvania, this in two directions.

Mr. Glick contributed some fascinating information and much of the section on Conestoga wagons, and gave permission in his official capacity to reproduce a series of photographs illustrating the development of harvesting equipment and machinery.

The writer is further grateful to George A. Wolfe, Ph.D., Director of Recreation and Facility Management, and to Bess Grace, Program Coordinator, BOB EVANS FARM, Rio Grande, Ohio, for permission to use photos of historic objects in the Bob Evans Farm Museum, and for authenticating information on the objects. Appreciation also goes to Tom Tilson for photographs of pieces in the Tilson Collection of farm-related collectibles.

Special thanks to David O. Percy, Ph.D., Director of The National Colonial Farm of the Accokeek Foundation, Accokeek, Maryland, for permission to use photographs, and to the photographer, Mrs. Kathleen Carlson, Farm Staff Member.

The National Colonial Farm, located in Maryland directly across the Potomac River from Mount Vernon, is a working farm, typical in its crops and animals of a middle-class tobacco plantation of the mid-18th century. Season farm work and domestic chores go on each day, and on weekends there are special demonstrations which focus on a particular aspect of colonial life.

With much gratitude, the writer acknowledges the help given by John Rice Irwin, Owner-Operator of the Museum of Appalachia, Norris, Tennessee, for permission to photograph and reproduce some of the many thousand items of the Museum's extensive authentic collections. Set up as a living mountain village of the Southern Appalachians, the Museum has many agriculture-related pieces from pioneer and frontier days. Mr. Iwrin aided also in answering a number of questions about items unique to his geographic area.

Personal thanks go to Howard Lynch, Fairfield Antiques, Lancaster, Ohio, for permission to photograph farm items, and for answering queries about objects I sometimes could not immediately identify.

Thanks go also to my father, Luther C. Hothem, for information on heritage antiques down through the years, and to William C. and Virginia McClurg, for assistance on this and other projects.

Special appreciation, finally, to my wife, Sue McClurg Hothem, for continued encouragement and support while researching and writing this book, and for all the small touches that have made this writing life both enjoyable and meaningful.

Lar Hothem

Introduction

Whether at obscure country auctions or in the better antiques shops, a broad class of items increasingly attracts collector interest. These are the farm antiques and collectibles. All are related in some way to rural life and the heritage of farming people, their barns, outbuildings and fields, their tools large and small.

Many thousand different objects can be associated with rural America. They begin with handmade primitives in the days of log cabins and a clearing in the forest. They continue to the more comprehensive multi-acred and well-tilled croplands of the much later agri-business when farm items were largely machine-made.

Most of the pieces sought today can be encompassed, timewise, in the broad span of about 1700 until about 1950. This offers both the aspiring and advanced collector two and a half centuries to browse through, selecting period or piece to suit checkbook and interest.

This book, in coverage, is somewhat oriented to the Midwest, for several reasons, though mention is made of collectibles from other regions. Midwestern farms have always had a broad mix of animals and crops, providing a spectrum of associated objects.

Too, the writer has lived on six Midwestern farms, and is thus somewhat familiar with the rural lifeway, what farms are about. From this experience a certain appreciation has developed for most objects associated with farming.

A growing number of people are collecting farm items whether individuals came from farms, villages or cities. Often, such solid memories from the rural past bring the best prices and are the best investments. At issue of course is monetary demand based on quality.

Farm goods rate highly, in that they were never really meant to be just collected. They were made and purchased because they were well-designed and well-made. Utility in turn depended upon strength and simplicity, the ability to do a certain task for years without repair or replacement. This really is one of the attractions of farm items, beyond the fact that most are no longer in use today.

A further inducement to the buyer-collector is surely the association of these objects to times that were simpler, more physical and direct. Nostalgia has limitations, but people were once proud of being fine rail-splitters and fence-builders. They were pleased with weedless gardens and arrow-straight rows of corn.

Once rural people did almost entirely for themselves, whether to provide shelter, transportation, or entertainment. Perhaps, as if by magic, we can gain something we need now by possessing things other people needed to exist then. All these are healthy aspects of this vigorous collecting of farm-related items, now a growing and expanding collecting field.

Author's Note

The book material is presented in a straightforward manner. Each category or area of farm antiques or collectibles is first *identified* by name, and listed alphabetically. Where pertinent, reference is made to related fields. This is immediately followed by *value* information.

Identification portions are written in an article format, listing a number of items in that particular collecting field. Background information and further facts are included whenever possible to highlight the importance of the item(s) in farming life.

Values here depart from the ususal practice of listing a single, specific worth, and a range has been put down. A few words on the use, and meaning, of this range might be helpful.

Generally, the lower figures represent a known auction bid or dealer's price for the item described. The higher figures represent a reasoned guess as to the maximum such an item might typically sell for or be valued at. A value about midway between the two might approximate a fair market value.

However, many other factors weight any final, exact values—including how badly a bidder or buyer wants the piece. Prices vary in different parts of the country, with higher values generally on and near the East and West Coasts. (New York and California are frequently cited, sometimes even by people who know.) And since many farm items could be purchased ready-made or be homemade, there are extreme fluctuations in quality and design, workmanship and materials. As in any collecting field, physical condition remains all-important.

As many of the major farm-related collecting areas as possible are covered in these pages. For subsequent editions—if a reader has a collection of farm items not mentioned or pictured here, and would be willing to contribute information and photographs—the author may be contacted in care of the publisher.

Lar Hothem

Adzes

A good dictionary refers to the adz as "an axlike tool with blade set at right angles, used for dressing wood". All sorts of adzes turn up at farm auctions, from railroad to shipbuilders'. Railroad adzes often have a double blade at opposite sides of the head, and some had an edge for rough cutting, one for smooth.

Ship adzes can be encountered, and the poll opposite the blade has a short, blunt spike. This was used for driving spikes into planking so they would not damage the blade. The average adz to be found is the general-purpose or "carpenter's" adz, designed mainly for working wood. It's poll has a large, flat (sometimes square) pounding surface.

This adz, and its relatives, was used to smooth beams, even early floors, cutting away the surface distortions left by the broadaxe. The marks left on the timbers of old barns are often those of the adz.

These came in two sizes, long-handled (used with both hands) and short, used with one or two. These are popularly known as "cooper's adzes", for the builder of kegs and barrels did use them. However, they were also general-purpose tools, swung when close and careful facing was desired.

Values

Adz, old handle, good condition, 8 in. blade	$25-35
Adz, cooper-type hand style, 10 in. handle	$25-30
Adz, coopers, "D. R. Barton/1892", 9 in. blade	$30-40
Adz, 5 in. blade, 33 in. handle	$35-45
Adz, small, used for making early dugout canoes	$40-50
Adz, straight-edge blade, 2½ in. across, handled	$30-40

Animal Carts
(See also Carts - Farm)

Wanting to have more fun and feel as grownup as adults, several animal-pulled carts were developed for children. These were derivations of horse-pulled vehicles and part of their charm today is that very few examples exist. They seem to have been manufactured in fairly small numbers in the first place.

An exception is the pony cart, an about two-thirds sized copy of a stripped-down horse-drawn surrey or road-cart. Another type resembled an early farm cart and had a high, box-like bed. These had play value for younger children, some utility purposes for the older.

The dog- or goat-pulled carts were strictly for play, and each was often complete with down-sized sets of leather harness. Some were factory- or custom-made to be exact replicas of then-current conveyances, while others were merely wheeled seats good for a fast ride.

The pony-cart probably survives in the greatest numbers, while there is some problem with the much smaller carts pulled by dogs or goats. These tend to merge as to size, and it's difficult today to say for certain whether a particular example was made for a large dog or small goat. All have twin tongues or side-shafts and light frame bodies.

Values

Pony-cart, surrey/road-cart style, 30-in. wide seat $80-100
Animal-cart, replica of farm wagon, traces of gilded
    trim, about 4 ft. long $230-250

Apple-Related
(See also Orchard Items)

Apples were by far the most important fruit on farms, and if any trees were planted, they were first of the large apple family, costing little, giving much. This was one food that could be enjoyed, in whatever way, year-round.

Apple-processing tools are common, some antique models still being used by back-to-the-land people. Apple devices can be logically gathered into paring or peeling categories, plus coring and quartering.

The first are the early all-wood arrangements for removing the skin, most crank-turned, some with a leather-belt drive. Some home-made, some store-bought, all had a revolving blade or turned the apple to the blade-edge to peel away the outer covering. Multi-purpose hand tools were popular, those for both peeling and coring individual apples. Some "gang" peelers could handle several apples at a time.

Skins intact, apples could then be preserved whole by hanging, or storing in sand or sawdust. Others were canned as slices, or further reduced for pies or drying. Coring machines were used, and most types punched out the apple center with a circular protrusion. American expertise also provided the unit with four blades, so the apple was divided into that many sections at the same time.

A number of hand implements relate to making applesauce and applebutter, the last a rich, dark preserve still put up in some communities and on large farms today.

Cider presses were on the earliest farms, though cider-making was then almost a profession. The word is incorrect, however, in that the presses only made apple juice; time made it into cider. A conscientious press operator put in a selected mix of picked and "windfall" or dropped apples, using the bruised fruit for body and taste.

The first gravity or screw presses were huge, the size of small buildings, often roofed. Each press run, powered by people or horses, produced barrels of juice from the pomace or "apple cheese". Eventually, for self-sufficiency, farms made or purchased smaller presses geared to orchard size.

Apples were ground into pieces and the pulp was then pressed, often one cider-mill doing both tasks. One end had a crank-action grinder, while a separate part held the screw-down wheel or opposing arms lever arrangement for compression. Eventually the liquid became sweet, then hard, cider.

Probably more than half of all orchard collectibles deal in some manner with apples.

Values

Applebutter stirrer, long-handled, wooden, 89 in.
   long $35-40

Apple-peeler, all wood, homemade, mortised construction, on 8 x 14½ in. pine plank, 8½ in. high $80-95

Apple-corer, tin, cylindrical, 7 in. long $10-13

Apple-corer, tin, hollow handle, 6 in. long $9-12

Hatchet and Hatchet head.

Top, broad-hatchet, factory-made, iron head, old handle. $16-20

Bottom, hand-forged head, old. $9-12

Broad-axe and broad-hatchet, showing relative head sizes; axe blade is just over a foot long. Each tool has an old if not original handle.

Broad-hatchet $19-25
Broad-axe $55-70

Upper
Adz, cooper's, 8 in. handle, marked "2" and "B" in a circle.
$35-42

Lower
Lathing hatchet, "C. Hammond / PHILA." 13 in. long. $10-15

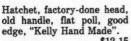

Hatchet-head, "Outeasy", 5 in. high.
$4-7

Hatchet, factory-done head, old handle, flat poll, good edge, "Kelly Hand Made".
$12-15

Apple parer or peeler, "White Mountain /Made By Goodell Co." Has "knocker" to remove peeled apple, piece 13 in. long. $25-30

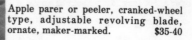

Apple parer or peeler, cranked-wheel type, adjustable revolving blade, ornate, maker-marked. $35-40

Augers, drills and assorted boring equipment. At lower center right is a rare pump-drill.

Photo courtesy Museum of Appalachia, Norris, Tenn.

Apple-peeler, mechanical, "Goodnell Co.", direct-
pivot model $55-65
Apple basket, gathering, carved double handles, 11½ in.
high $95-115
Apple basket, nailed rim, wire-bail handles, 7 in.
high $12-15
Apple basket, handmade notched handles, half-bushel
size $90-110
Apple basket, dark splint, New Hampshire type, 20 in.
diameter $115-130
Apple peeler, cast iron, "Sinclair Scott Co./Baltimore",
8 in. high $33-45
Apple peeler, cast iron, "R. P. Scott & Co./Newark,
N.J.", 8 in. high $42.5-55
Applebutter stirring stick, 36 in. long $10-15
Apple peeler, "Victor", cast iron, 9 in. high $35-45
Apple peeler, wooden flywheel, primitive, 32 in. long . . $40-55
Jar, applebutter, redware, interior glaze, 8¾ in. high . . $85-100
Cider press, "Hocking Valley", with grinder and press
together, wood construction $50-65
Cider press, hardwood slatted press chamber, side grinder,
41 in. high $95-115
Cider jug, stoneware, one gal. size, white top, brown
base $22-26
Apple box, square, wooden, orig. paint, 9 in. square,
2½ in. high $35-45

Augers

Tools for making or enlarging holes were necessary for a large
number of farm enterprises, from maple syrup to erecting barns.
They drilled holes for the sap-run and peg-holes in building
frames. Augers differ from hand-drills and drilling machines in
that they tend to be older and simpler.

A common auger form has a thread diameter of over an inch,
and an iron shaft secured to a wooden crosspiece handle. They
relied on muscle power and a sharp eye for "heading" a true
hole, both angle and distance. Two hands were required, but
small "gimlet" augers were also used with finger power alone.
One of the most unusual farm tools ever made was the pipe

auger, out of use for over a century. Before the days of clay and castiron pipes, oak or other hardwood logs occasionally served as pipes for running fresh water. Laid underground and connected together, they were the first water-lines. The long log auger bored the central hole, this 1 to 4 inches in diameter.

Another tool with one application was the bung-auger, which made a keg or barrel useful. It shaped and enlarged the bung-hole in a stave so that contents could go in or out. Bung-augers are conical, with boring tip and scraping edge, and wooden twist-type handles.

Values

Auger, wood handle with central brass casting, 27 in.
    long $40-50

Auger, wood handle, good condition, 14 in. long $13-18

Auger, gimlet, all metal, 6 in. long $4-6

Auger, pipe, shaft over 5 ft. long, no recent examples
    known to have been sold . . . .

Auger, bung, 9 in. long, wood handles, good blade $30-40

Auger, hollow, "Stearns & Co.", 6¾ in. long $22-30

Auger, hollow, "The A.A. Woods & Sons Co.", 6½ in.
    long $25-30

Axes

Many travelers and observers in the 1650-1850 period have remarked on the trees of North America. This is especially true for the region east of the Great Plains, all the way to the Atlantic Coast. There are reports of giant hardwood stands so thick that one could not walk a dozen paces without encountering a trunk. There is a journal entry by one man that he suffered severe depression, because for days on end he could not see the open sky.

Such was the story in many areas when early pioneers entered the Eastern Woodlands. They came to carve out, if not a personal empire, at least a homestead. The land was taken by grant or purchase, but it was held and tamed with the axe. It may well be that the axe was the single most important tool of early America.

Actually, there was no one tool known as "the axe". There were several kinds, each designed for a different job. Of these old axes, there are two classes, neither of which could easily do the work of the other.

First was the felling (falling) axe. It's name derives from the function, which was felling or cutting down trees. This was the axe used to clear forests for fields, to outline the owner's land. Further, it cut up the tree into manageable lengths, and lopped off branches. It was the standard firewood axe when that material was the only available energy source.

Next was the broad axe, so-called from its broad, lower blade. It could possibly have been used to cut down timber, but it would have been clumsy and inefficient. Instead, it was used for shaping the fallen tree, so that it could be used for a log cabin or for the interior structural timbers of a barn or shed. In the basement of very old homes, the beams often bear the marks of the broad axe.

It is not too inaccurate to say that felling axes cut down the forests, while broad axes put up the dwellings. As to which was most important on early farms, it would depend on what was needed most at the time—a field or pasture, or a roof for the night.

Very early felling axes from the 1600's have one curved blade, and a round or oval eye for the haft or handle. The first axes in America were patterned after European utility tools and battle-field axes. Most of the these were heavy for day-to-day activities, and lacked the sturdiness of later American axes.

Both felling and broad axes were heavy-duty tools. The felling axe was 6 to 8 inches high, had a squarish head, and was fully developed by about 1800 in this "American style". The American broad axe was 8 to 10 inches high, with the blade length averaging a foot or so, often longer. In weight, the iron head of the broad axe was more than double that of the felling axe.

And while the felling axe resembled modern single-bit axes, the broad axe had a peculiar shape. It was, for the most part, more difficult to make, and not just because it was larger and heavier. One side had to be almost flat, for this was the portion used against the wood surface. The other side was beveled in different ways, so that the iron combined strength and lighter weight.

The cutting edge was either straight or gently up-curved at the ends. The width compared with the blade edge of the felling axe was much greater, as the latter was 4 to 5 inches. In use, the broad axe made regular cuts in the fallen tree, along one of the four sides, and to a depth of 1 to 3 inches. Such cuts were made 3 or 4 to the yard. Then, the in-between sections were chopped out, and before long a round trunk section had became a square beam.

Besides the size and shape of the iron heads, the axes had different handles, and they were also swung in dissimilar ways. A felling axe, with symmetrical head, had a fairly straight handle some 3 feet long. It was swung from the side, biting into the trunk at angles that averaged 90 degrees.

A typical broad axe had a 2-foot handle, though tall men preferred longer ones. The axe was used with an up-and-down motion, almost like hoeing weeds, to first score a log. Most broad axe handles were sharply canted toward the flat head side. This permitted the user to stand on, or straddle, a log and still work the sides. Most broad axes were hafted for right-handers, but a few left-hand examples exist.

Both axe types were made of varied-quality irons, and were either cast or wrought. A cast head (generally after about 1830) was formed when molten iron was put into a mold. A forged or wrought head was created on the anvil of a blacksmith's shop. Another way to say this basic distinction is that one axe was poured, the other pounded, into shape.

Felling axes were of either manufacture method, for the simple, even dimensions permitted mold use. Broad axes were generally made up in steps, for they are complex objects. The wide lower blade is relatively thin, as is the iron around the eye.

Both axes would have been almost useless if they had been made entirely of iron. Steel was required for the cutting edge, and this was achieved in two ways.

Cast felling axes had the blade edge carefully tempered until it became mild steel. It would then take and hold a sharp edge. Wrought broad axes, when being made up, had a long steel insert put in place, while other types had an overlaid steel piece. This was then welded into a permanent bond. Both methods of manufacture required final hand-filing to remove burrs and make an even blade edge.

Early farm felling axes did not have a pounding poll above the eye, opposite the blade. Pioneers could cut with it, but could not pound. As the axe developed in America, it acquired a rounded or squarish mass above the eye.

This not only strengthened the axe head, but allowed the axe to be used as a light sledge-hammer, so it became a multi-purpose tool. It could now drive wedges to split logs into firewood or could pound the long pegs or trunnels (tree-nails) into place, those that held barn and shed timbers together.

Beyond the general axes (and felling axes later acquired double edges for Northern timbering) there were other, special-purpose types. The 18th Century goosewing hewing axe was a type of broad axe that had a long blade extension. This projection from below the eye went either beyond the eye, or back beneath the handle, but not both. The regular broad axe has the lower blade extending about equidistant from the eye.

The goosewing was much-used wherever heavy but well-controlled cuts were needed. The goosewing was made in a huge variety of styles, and the more artistic types have been called "angel wing".

Mortising axes had a long, very narrow blade, generally sharpened on both blade edges, not just on one side as with broad axes. Sometimes also called post-holes axes, they made the deep openings in fenceposts that received the ends of wooden rails. They were also used to hole large timbers for barn frames.

Turf axes had very large, round or rectangular lower blades. Despite the size, they were rather lightweight. Most are believed to have been used in New England to cut sod or peat, to enlarge drainage ditches or shape sod blocks for building walls.

The lore of the axe has given us a number of colorful terms, ones we use daily. "Fair and square" may refer to the work of the broad axe when skillfully done. "Flying off the handle" meant the iron axehead whirled away due to improper haft fitting or wedging. "One fell swoop" means the last blow of the axe which starts the tree earthward.

Collectors look for dated axeheads, those "signed" with maker's mark and/or location of maker. Some axes had the owner's name burned into the handle. A few rare axe-handle forms still turn up, the original, personalized handle from which new ones were copied.

Values

Axe, double-blade, head only, "Sager", 12½ in. long ..	$15-20
Axe, mortise, for holing beams, 35 in. long	$75-90
Axe, felling, wrought head, ca. early 1800's, 6 in. high	$40-50
Axe, broad, steel edge, 13½ in. long blade	$50-60
Axe, broad, "D. R. Barton/Rochester, #2", old handle	$75-90
Axe, broad, head only, edge 12½ in. long	$20-25
Axe, broad, old slanted handle, dated 1832	$75-85
Axe, broad, New York, 13 in. blade edge	$45-55
Axe, broad, with handle, "Special/IXL", good condition	$65-80
Axe, broad, 30 in. handle, "J. Godfrey/Warms"	$75-90
Hatchet, broad, head only.....................	$8-10
Hatchet, broad, handle 16 in. long..................	$13-17.5
Hatchet, broad, marked blade, label on handle	$20-25
Axe, goosewing, no marks, old handle, edge 14 in. long	$175-225
Axe, turf, handle 30 in. long, from New England	$30-35
Axehandle form, never had head, wood, with hanging hole	$20-25

Barbed Wire

Barbed wire or "bob wire" was widely introduced by the third quarter of the 1800's. A heavily collected field today, barbed wire was once the cheaper standin for wood or stone fences. And, per rod, it was easier to obtain and put up.

In the East wire was used as much to protect crops as fence animals. In the West, it was used to protect railroad right-of-ways, fence waterholes and boundry private landholdings. Barbed wire figured in range wars for water and grazing rights, cattle and sheep conflicts, and marked in slow stages the end of the open government range.

The main genius of barbed wire—perhaps a thousand

different types were put out in the past century and more—was the protruding points that caused farm animals to shy away before the fence broke.

Collectors accumulate barbed wire in 18-inch lengths, and a few have hundreds of different styles, many identified and dated. Those strands with only light rust, all wires unbroken and original barbs correctly spaced are preferred.

Values

Strand, three-point, double wire, rusted $3-4

Strand, pointed galvanized discs, single wire $18-24

Strand, four-point with double duo-tip wraps $7-10

Strand, "high visibility", with wooden blocks between
    barbs $30-38

Baskets

Baskets were put to dozens of uses on the family farm, from picking orchard produce to plucking goose down. The containers were used for picking strawberries and picking up potatoes, for gathering eggs and vegetables.

The majority of baskets were made from thin wood splits or splints, and certain oaks and hickory were the favored woods. Other materials were used, including willow strands, whole or split, and other peeled twigs. Lighter substances like ryestraw, swamp cane and sweetgrass were used as decoration or the entire basket body.

While most farm basketry, especially in pioneer times, received rough and regular usage, much damage occurred due to rodent holes or improper storage. And, relatively speaking, though baskets are quite strong for carrying or storage purposes, they were still vulnerable to misuse, fire and the elements. So it is that top-condition baskets are the most scarce and command the best prices.

Basket-collecting is a major field today, and the demand has forced prices up sharply in the last few years.

Values

Gathering basket, mustard paint, 12 in. diameter,
    wood $50-60

Gleaner's basket, woven splint, 31x45 inches.......... $100-125

"Go-to-market" basket, rigid handle, unwrapped rim,
8x16¼ inches $75-90

Market basket, nailed rim, arched bentwood handle,
23½ in. long...................... $55-65

Melon-type basket, twisted peeled-twig handle, 22 in.
long $110-125

Produce basket, elongated, splint, 18 in. long $55-65

Sifter basket, splint, bottom holes, 8½x13 inches $65-75

Basket, gathering, woven splint, wood side handles,
rim damage, 30 in. square.................... $25-35

Basket, woven reed and splint, wood bottom, side
handles, 12x20½x29 in. long $45-65

Basket, twin-bottom "buttocks" type, 5½ in. high $40-50

Basket, side-type, single-bottom, splint, 11 in. long $95-120

Basket, pea-picking, wood, 20 in. long $62.5-75

Basket, picket-fence type, wood bottom, flared sides,
wire rim, 8x 11¼x18½ inches $75-90

Basket, splint, handholds, 15x23x33 inches $65-85

Basket, wood splint, for goose-down collection, lidded,
25 in. high, 14 in. square at base, round top $185-225

Basket, twin-bottomed, wood handle, 17 in. diameter.. $90-125

Basket, splint, nice handle, body 5 in. high $65-85

Basket, round top, square bottom, wood splint, 15½ in.
diameter $45-65

Basket, splint, potato-print designs, 15 in. diameter.... $70-90

Wicker basket, oval, 9x12½x17 inches............... $49-59

Wicker basket, lidded, handles, 15x20 inches $53-63

Winnowing basket, woven splint, 16½x18 inches $32-42

Wood basket, splint, good old color, 10x21x26½
inches $80-95

Woven splint basket, minor damage, 24½x37½ inches . $135-150

Woven splint basket, 8x10¾x18 inches $37-47

Bee-Keeping

Domestic honeybees were brought to the U.S. East Coast from
Europe in 1638 and were taken to the West Coast (New York to
California) in 1853. Since those years, bee-keeping has

flourished. Their greatest commercial values goes beyond honey and wax, for they are vital to pollinating plants and trees. A number of collectibles are related to this somewhat esoteric occupation.

Perhaps the earliest is the bee-box, often a handcarved wooden container in which captured bees were retained so they could be followed in their bee-line flight back to the hive or honey tree. Powdered chalk was sometimes used to mark the bees.

Later, but still early is the bee *skep* or hive for domestic bees. It is conical, a bit less than 2 feet high, and most were made of twisted and wrapped ryestraw. Pennsylvania may be the source of many such examples, though they were common in New England. Few are in good condition due to age and the elements.

In more recent times, bee-keeping was part of every large orchard and many general-purpose farms, those combining crop and stock raising. To obtain the honey, a bellows-type smoker was used, which stupified the golden insects to prevent stings. Some smokers operated with a crank handle.

During cold months, bee feeders were hung in the hive to keep the occupants alive, while special large brushes were used to clean insects from honeycomb frames when the hives were "worked".

Special containers held queen bees when part of the hive was ready to swarm or fly off to establish a new hive. As long as the keeper had the queen, the new swarm would remain close to her and could be transferred.

If a swarm did get away and formed a mass around a tree branch or under the eaves of a building, a knowledgeable person could use a funnel-like screen trap mounted on a long pole; the trap had a tight-fitting cover. It had an appropriate name, the "swarm-catcher".

Associated bee collectibles would be the various containers, the jars and jugs in which honey was stored. At first, glass jars were never used, even when available, as it was thought that light ruined the quality of honey. Only recently was honey put up in small, clear containers. Wax was an important by-product, used to dress leather, and in the late 1800's sold for 37½ cents a pound.

Values

Bee-box, sliding lid, airholes, 3½ in. long $35-40

Bee-box, pine, old square handmade nails, tin exit port . $45-55

Skep, conical hive, bound ryestraw, weathered, 18 in.
    high $50-65

Smoker, tin, leather bellows, handled $11-15

Smoker, sheet copper, fabric bellows, copper spout $45-55

Feeder (called "honey board"), wood slats, for eight-
    frame hive $3-4

Bee brush, wood handle, 16 in. long $4-6

Queen bee carrier, tin, cylindrical, swivel cover, 3 in.
    diameter $85-100

Honey extractor, hand-cranked, frame with spinning
    racks to remove honey by centrifugal force, 17 in.
    high $45-55

Swarm-catcher, conical screen, without pole, about
    18 in. high, flap cover $20-25

Honey jar or jug, tan stoneware, bail handle, lidded,
    narrow, curved pouring spout, 9 in. high $24-28

Honey tin, square, half-gallon size, two top openings
    for filling and pouring, old paper label for orchard $28-34

Bells

    At mention of farm bells, two types come easily to mind, one
gastronomic, one nostalgic. They are the dinner bell and sleighbells. The former called in family and hired hands for dinner or
supper, and the latter provided a musical accompaniment to a
winter journey.

    Dinner bells are of two widely different kinds. Larger farms
had the pole-mounted cast-iron bell, with heavy tongue or
clapper, these ranging in size from 8 to 15 in. at bell-mouth
diameter. The bell was sounded either by activating a cord or wire
attached to the clapper, or by a system that swung the bell itself
to and fro, making the bell sides strike the clapper.

    Hand bells were used as well, these being larger sizes of the
schoolbell variety. Of cast brass, they have natural or japanned
(black) wooden handles and cast-iron clapper. Bell-mouth
diameter is 5 or 6 or more inches, and such bells are also known

Basket, early factory product, handles at each rim end, nailed and woven construction; basket 20 in. wide. $30-40

Burl-headed mallet, hardwood handle, about 12 in. high.

$23-28

Bootjack made from circular-saw cut plank about 1 in. thick. $9-13

Fleam, brass case, unmarked, straight-edge blade and three sizes of blood-letting blades, about 4 in. long. $60-75

Twin-bottom basket, 9 in. high to handle top, lower circumference of 34 inches, excellent condition. $110-140

Twin-bottom basket, 17¼ in. to handle top, lower circumference of 59½ inches. In unused condition. $190-230

Left, rope-making or strand-twisting machine; right, farm-made brooms.
Photo courtesy Museum of Appalachia, Norris, Tennessee

Bee-smoker with attached bellows, tin, 12 in. high.

$19-26

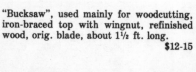

"Bucksaw", used mainly for woodcutting,
iron-braced top with wingnut, refinished
wood, orig. blade, about 1½ ft. long.

$12-15

Brace for taking bits for drilling
small to medium-size holes,
hardwood handle and rest, 14
in. long. $8-11

as "huckster's" from the days they were rung from horse-drawn sales wagons to announce an arrival. Both farm bell types were foundry cast.

Sleighbells were attached to the harness and movement of the horses caused the bells to jingle merrily when the sleigh was in motion. The bells were secured in graduated sizes to long harness strips, each size giving a different pitch. Some strands had many dozen bells per matching set. The main purpose of the bells—besides delighting both riders and the occupants of nearby homes—was for night warning of an approaching sleigh.

Cowbells are large animal bells, and some of the first ones were hand-carved of hardwood. Later iron bells were handmade by blacksmiths until factory examples became common. The regular mellow strikes had a calming effect on cattle.

Besides the tall, rectangular sheet iron cowbells, small cast-iron and cast-brass animal bells were often used. These had a top-loop for attachment to neck straps, and from the smaller size it is obvious many were used on calves, sheep and goats. Overall, such bells let the farmer know where the animal was, and whether it was content or in distress.

The smallest farm animal bells are the turkey bells, with mouth diameter of 1 to 2 inches. An early ad for a turkey bell states, " . . . enables the flock to be easily located, makes the foxes shy".

Values

Large cast-iron farm bell, post-mounted	$70-90
Large cast-brass hand bell, huckster-type, wood handle	$85-130
Sleighbells (example, 36-bell set, graduated size, black leather in good condition)	$100-150
Cowbell, handmade, hardwood, wood clapper, 7 in. high	$35-50
Cowbell, handmade, sheet iron, 5 in. high (higher value if with maker's mark of identifiable blacksmith) . . .	$12-18
Animal bell, factory cast-iron, 3 in. bell-mouth diameter	$15-20
Animal bell, factory cast-brass, 4 in. bell-mouth diameter	$35-40

Turkey bell, factory cast-brass, 1¼ in. bell-mouth
diameter $15-20
Note: Most animal bells are valued at several more dollars if the
original (or an old) leather strap is present.

Benches

Benches, for sitting on or setting things on, were common on
farms. Characteristically long and low, simple and without
backs, they were located at strategic places. One was generally on
or near the back porch of the farmhouse, in front of the milkhouse, or in or near the out-kitchen.

Often, a long bench was set under the grape arbor, even
attached and permanently fixed to supporting posts. This
provided a cool resting place in hot weather.

Single-board tops were the rule for most farm benches, and
sturdy end-supports, perhaps with scroll cut-outs at the base.
Most benches had brace-supports to prevent lateral swaying, or
wide runners beneath both edges of the top or seat secured to the
uprights. Pine was a favorite material, though woods like poplar
were sometimes used.

Values

Bench, legs mortised into top, old blue paint, 17¾ in.
high $235-265
Bench, cut-out legs mortised through top, 12x87x18 in.
high $95-120
Bench, legs mortised through top, modern white paint,
11¾x48½x15 in. high $40-50
Bench, splayed legs and old grey paint, 49x19 in. high .. $50-60
Bench, worn old dark green paint, 9½x70x17 in. high .. $95-120

Berry Picking

No longer of major importance, picking wild berries was once
a valued family enterprise, at least while such natural gifts were
ripe. This was the special realm of farm children, though adults
often joined to increase the yield.

Berries were valued, in addition to off-the-bush treats for
pickers, for two reasons. They provided sweet preserves (jams

and jellies) for the cold months, and were a cash source for the youngsters who could sell door-to-door or set up a small roadside stand.

Going "berrying" meant long hours of tedious labor, the stamina to ignore mosquitos and stinging flies, and the good luck not to kick over the "keeper" or storage basket. Farm children had a large selection of wild or tame-gone-wild berries to choose from, and most berries timed themselves nicely to ripen at different seasons.

Earliest were the varieties of strawberries, followed by red and black raspberries, then blueberries (or huckleberries). Some folks also picked elderberries, which could be made into pies or the semi-poisonous polkberries which would produce a fine purple ink. Ground-vine delights like dewberries were also sought in some areas. Almost everywhere, blackberries were picked in great quantities, to be made into pies, preserves, and even wine.

Values

Berry basket, tin, old, wire handle, one-quart size $15-25
Berry basket, woven wood splint, handmade, 6 in. high $45-65
Berry basket, factory-made, wood splint, 5 in. diameter $12-25
Berry storage basket, wood splint, 14 in. high $60-80
Berry tin, bail handle, copper, one qt. size $50-65

Birdhouses

Farms provided homes for most beneficial creatures, from cats to dogs, and "good' birds were no exception. House-sparrows, pigeons and starlings made themselves comfortable, unwelcome guests that stayed.

Birdhouses—even feeders and baths—are collectible, providing they have a certain attractive quality and some unusual features. Species coaxed to stay close included bluebirds, martins and wrens. The first were little loners that needed solitary and secure nest-boxes, while wrens preferred accomodations near people. Martins, the fleet communal flocks, like apartment-style living far off the ground.

Friendly birds were encouraged for three reasons. Their songs were pleasing and their antics amuzed the household. The farmer was well aware that the birds devoured huge quantities of harmful insects, and more than paid their way. Astute observers could even predict weather by bird behavior.

Birdhouses were generally hung from the grape arbor or a tree limb, though martin colonies always preferred pole-mount. Wood was the common birdhouse material, and scrap boards were often used.

Rarely, the birdhouses copied the farmhouse style. Tin and plank-built examples can often be found, as well as the occasional makeshift home made from a packing crate. One rare wren's nest was made of pottery, complete with entry and perch-hole which once held a twig.

Values

Birdhouse, hanging, wired, small staved keg with end-
    hole, old, weathered, 9 in. long $20-25

Birdhouse, hanging, homemade, cottage-shape, old
    paint, good condition, 6x10x8½ inches $8-12

Birdhouse, pole-mounted, made as two-storey home
    with four nest-box areas, 14x16x15 in. high $17.5-25

Birdhouse, wren's, hanging, old wire, cream-glazed
    pottery, round, 11 in. high $75-100

Blasting

On May 8, 1903 (according to an old ledger) a John Bashore in Pennsylvania purchased the following: 11¾ lb. dynamite, 5 yd. "fues" (fuze) and 10 blasting caps. In so-doing he typlified the farm tradition of always improving the land. Usual blasting targets were stumps and large rocks, with roadbuilding a close third. Even small, meandering streams were straightened by blasting to lessen flooding problems.

Until the late 1800's, blackpowder was the chief commercial explosive available, fired by match and fuze. Then, Alfred Nobel's dynamite came into wide acceptance, and it proved superior as it was more compact, powerful, and safer.

A fuze led to a heat-fired blasting cap, itself inserted into a dynamite stick. There are accounts of first-time dynamite use in which ambitious amounts were put together to blow stumps, with a new problem—filling in the resultant crater.

Values

Keg, wood-staved, wood-strip lapped, for bulk black-powder, 12 in. high	$75-95
Powder box, Hercules, factory shipment container	$30-40
Powder box, Du Pont, for factory shipment	$30-40
Fuze box, heavy cardboard, 10 in. long, 1 min. delay	$3-5
Cap container, tin, faint markings, 3 in. diameter	$6-8

Bleeders

Before "animal doctoring" developed into the veterinary profession, farmers made up their own medicines and treated livestock themselves. One necessity was the bleeder or "fleam", an instrument designed for blood-letting. Science of the times ascribed many ills, human and animal, to "bad blood", and misguided practioners regularly bled both ill people and farm stock.

Some bleeders are simple sets of several blades and lancets folded into a brass case like penknife blades. Most have graduated-size projections, razor-sharp, at blade ends for cutting veins, and all originally had a case of wood, leather, or even papier-mache.

Other bleeders were spring-driven and cut when a trigger was pushed or pulled. Most extant examples seem to date from the early 1800's, and almost all are very well made from the best available materials.

Values

Bleeder, three projecting blades and a straight-edge, steel, in brass case, unmarked	$65-80
Bleeder, brass and steel, 3½ in. long	$85-100
Bleeder, brass and steel, in worn wood and leather case, 3½ in. long	$70-85
Bleeder, brass and steel, small, "Wilgand & Snowden/ Phila.", fitted case, 2⅛ in. long	$135-150

Blowtorches

When gasoline-powered vehicles and engines arrived in quantity on the farm, the fuel was also used in the hand-held blowtorch. With metal parts and body of extruded brass, the blowtorch—when the firing tunnel was preheated and the reservoir pressurized with pumped air—produced a steady blast of flame in the blue-hot range, around 1800 degrees.

Though sold mainly as a plumbing tool, blowtorches could also solder, weld and braze. Genius, however, was not confined to a college or laboratory, and inventive farmers quickly found a dozen uses for the sputtering beauties. They were used to thaw frozen water pipes, burn off brush and destroy yellow-jacket and hornet nests.

Blowtorches could, used carefully, peel off old paint, melt lead for casting bullets, and turn old lead pipe into new raw material. Some were even able to temper tool edges into new strength and longer life.

Values

Blowtorch, one-quart capacity, gutta-percha valve handle,
    polished brass, round body $22-27

Blowtorch, one-pint, rectangular with rounded edges,
    steel with copper-plating, early $24-29

Blowtorch, one-quart, lacquered brass base, brass turn-
    valve, maker-marked $27-32

Boot Jacks/Scrapers

Until the late 1800's, male farm workers all wore boots. They were the long, heavy "clod-hoppers" that reached to about the knee, and were sturdy if not pretty. While the boots pulled on easily with the aid of top loop straps, they came off with difficulty. Too, they were frequently mud-covered from field work.

Boot-scrapers were an important addition to any nearby area that required a clean floor, especially the back-door farm kitchen or the summer kitchen and milkhouse. Early boot-scrapers were simple wood or metal strips, usually secured to the floor, while later versions were store-bought. All were designed to scrape mud and dirt from the bottoms and sides of boots.

Once done, the still none-too-clean boots had to be removed. Farmers tended to get their feet dirty and keep hands clean, so the boot-jack was in high favor. Basically, this was a wood or metal piece that had a "V"-shaped opening to catch and hold bootheels as an aid in taking them off.

Values

Boot-scraper, plain, wood crosspiece, triangular edge . . $3-5

Boot-scraper, metal edges, horseshoe sides, wrought
    iron $15-20

Boot-scraper, low, wrought iron, 11½ in. long $30-45

Boot-scraper, wrought iron, for right angle of step,
    scrolled finials, 7 in. wide $55-70

Boot-jack, primitive, forked stick nail holes $7-10

Boot-jack, old, "V"-shaped indentation, raised end, cut
    from board $12-17

Boot-jack, cast-iron bug with projecting feelers $50-70

Boot-jack, cast-iron figure, "Naughty Nellie" $60-80

Boring Machines

Boring machines, with their assorted grouping of different-sized bits, replaced for some work two other common farm tools. They were the bit-and-brace, and the simpler hand-turned auger. The genius of the factory-made boring machine was two-fold; it allowed rigid, steady drilling, and, using the common two-handed energy drive, was quite efficient.

Machines that bored holes had metal-geared mechanisms within a heavy wooden frame which was designed so that the user could literally sit down at the job. His added weight on the frame extension allowed both secure and accurate hole-making. One major use was holing heavy timbers for pegging, as in building frame construction.

Many boring machines had the top portion secured so that it could drill at different angles, though some were designed only to bore at 90-degrees.

Values

Boring machine, with top framework able to angle two
    ways $65-80
Boring machine, "Millers Falls" maker, can be angled .. $75-90
Boring machine, stationary, 90-degree angle only $55-65

Bottles

At first glance, old bottles would hardly seem to be directly associated with the American farm, but they are indeed. From about the mid-1800's on, foods, animal cures, patent medicines, liquor, and sundry solids and liquids were sold in glass containers. Farm people purchased the goods, used the contents, and were left with the empty containers.

This was generally before the days of regular trash pickups. Glass containers, being nonbiodegradable and unburnable, accumulated. Even canning jars, when store-bought foods became cheap and safe, were surplus property.

As bottles of all kinds piled up, the waste required disposal. Four main depositories for these were used, and at least several could be found on any farm. Those farmers along waterways simply dumped trash near the bank and let the current and seasonal floods carry everything away.

Ravines were also used, and the head or beginning of small gulleys and arroyos all sparkled in the sunlight. Closer to the farm buildings, cisterns were used in later times. These were wide and shallow, stone-faced or brick-lined pits, underground reservoirs that had building roof runoff channeled into them. When the various handpumps over wells became popular, the empty cisterns remained an ideal dumping place, permanent, unseen.

Probably the best bottle-disposal site were outhouse pits, known to some present-day treasure hunters as "glory holes". this was the sanitary excavation beneath the structure. For whatever reasons, bottles of all types went into the pit, and such sites tend to have a mild depression in the ground.

Outhouses were moved every few years, and old farmsteads may have a number of them. One site, carefully excavated in Missouri, produced over 300 perfect specimens, some dating

Flashlight, "Eveready /Case no. 2612, 3-unit", metal fittings all plated brass, 10 in. long. $10-14

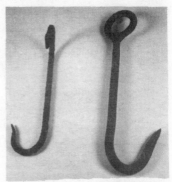

Meat-hooks, both hand-forged iron, left example in unusual form of snake, 6 in. high. $12-15

Right, heavier hanging carcasse hook, probably originally with chain. $7-9

Gambrels, assorted sizes and lengths, each $4-7
Keg, wooden, wire-strapped, probably for nails. $3-5

Butchering items; top, butcher-block scraper, 5 in. wide, wooden handle and steel blade. $4-6
Spouts for sausage stuffer, four different sizes, each of spun brass. Each $3-4

Meat chopper or grinder, cast-iron body, hardwood handle with brass sleeve, "Enterprise Mfg. Co.", about 13 in. high. $15-20

Butcher-block, on turned legs, solid hard-maple top, about 28 in. high. $150-175
Courtesy private collection.

Butchering implement, spatula, by "Lamson & Goodnow", brass rivets, handle impressed "Hommel" 14 in. long. $8-11

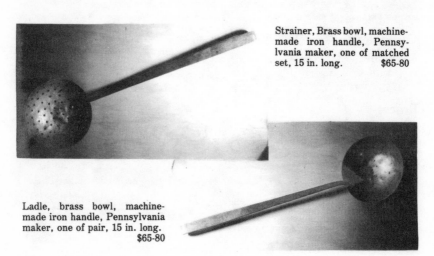

Strainer, Brass bowl, machine-made iron handle, Pennsylvania maker, one of matched set, 15 in. long. $65-80

Ladle, brass bowl, machine-made iron handle, Pennsylvania maker, one of pair, 15 in. long. $65-80

back to Civil War times.

Bottles can also be located in barns. Empty bottles are in out-of-the-way places, like under the first-storey groundboards. Many once held veterinary medicines and liniments for treating sick or injured animals. A great many farms still have accumulations of interesting old bottles found in the course of demolition and construction.

Values

Fruit jar, "Ball Ideal", pint, pale blue $16-20
Fruit jar, "Ball Ideal/Wire Side", pint, pale green $16-20
Fruit jar, "Atlas-Mason", one pint $9-12
Fruit jar, "Mason", two-quart, aqua $5-7
Medicine, "Nitrite of Magnesia", round, 6 in. high $45-55
Whisky, clear glass, oblong, man's face on wide side . . . $9-12
Soda, Coca Cola, area bottler, ca. 1940 . . . . $4-7
Flask, clear, "Great Seal", 4½ in. high $3-5
Decanter, stopper top, "Nujol", clear, 5½ in. high $8-10
Milk, half-pint "Hoak Dairy/San Bernardino", 4 in.
    high $2-3
Medicine, "Chas. H. Fletcher's Castoria", pale green . . $10-13
Bottle, "3-in-One Oil Co", ca. 1905, 4 in. high $7-10
Medicine, "Sloan's Liniment", clear, 5 in. high $4-6
Insect powder, cork top, "Black Flag", orig. label $3-5
Medicine, "C. H. Dunlap/Asthma Oil", clear, 6¼ in.
    high $7-10
Codfish oil, on side, fisherman carrying catch, brown . . $3-5
Medicine, "Tonsiline/For Sore Throat", giraffe's head
    and neck on front $9-12

Braces
(See also Boring Machines)

Braces, turning devices, were used with drilling bits to make medium-size holes. They improved on the small-hole (gimlet) and large-hole (auger) tools by the use of greater leverage and more accurate hole alignment. By the 1700's, the question-mark-shaped brace had become refined, and had a solid clamp-down or turnscrew to hold the bit firmly in place.

Early braces and a hand-tempered bit or two are far from common, and in collector demand. Easily found are the wide range of factory-made braces, most with metal-rod construction, metal screw-down bit holder, and wooden hand and palm grips. By the 1900's, complete kits, boxed, were available with a full range of nearly a dozen wood-boring bits. These typically ranged in diameter from ¼ to 1 inch.

Extra-fancy braces, highest of the brace-maker's art, were available, though the cabinet-maker, not the farmer, was the likely owner. Also, most hand tools, including braces, that were used by coach-makers were of the highest quality.

These were of exotic, hand-finished woods, with liberal use of polished brass parts and insets, and the use of hand-carved ivory was not unknown. These all are sought by advanced tool collectors, and sell for several hundred dollars each.

Values

Brace, angle attachment, "Millers Falls", 17 in. high . . . $20-25

Brace, brass ferrule on pad, 15 in. high $15-20

Brace and bit, fine wood, brass chuck and button, 18 in. high $70-85

Brace, chuck marked "Harrold/U.S.A.", 11 in. high . . . $22-26

Brace, screwdriving; "Drummond's Pat. 1870", brass cap on wood handle, 12 in. long $30-36

Branding Irons
(See also Stock Markers)

Most widely used in the West, branding irons had stock-marking heads with the rancher's or owner's personal mark. These were in a variety of imaginative shapes, all designed to leave a permanent mark on animal hide. Cattle were branded in annual drives or round-ups.

The purpose was to identify cows as to owner, and prevent rustling or theft by direct action or deception. So, brands were developed with marks that included or overlapped those of a neighbor so that part of his herd could be brought in. This was infrequently done, however, since divergence from the branding system was both monitored and had penalties.

Meat chopper, ca. 1850, with top removed to show interior; screw action, hand-cranked, graduated-space blades, 8½ x 10½ x 19 in. long. $90-125

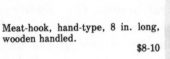

Cutting or chopping board with cleaver, board 14 in. long.
Cutting board, maple $8-10
Cleaver $9-12

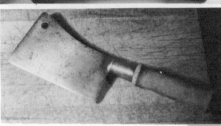

Meat-hook, hand-type, 8 in. long, wooden handled.
$8-10

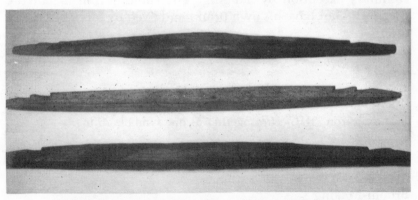

Gambrels, three sizes with middle example 28 in. long, handcarved hardwood, Midwestern. Each . . .
$5-7

Butchering items; top, factory-made iron meat-hook for storage or smokehouse. $1.5-2

Bottom, early Sheffield (England) general-purpose knife with bone handle, 6 in. long. $4-5

Kettle-scraper/stirrer for meat puddin', with handle about 4 ft. long. Marked on sleeve at end of wooden handle: "Sheble & Fisher". The factory was at Ashland and N. Fifth, Philadelphia, and this tool was made sometime between 1875 and 1885. $80-95

Beyond the alphabet and figural brands, the rarest brand form is the so-called "running iron". This was a straight, poker-like iron with tip that canted gently to one side. It was used to add to legitimate brands, or copy them freehand. It is rare because in some places in the West mere possession brought summary execution by hanging. The running iron meant the owner did not have his own brand and lived by forging those of others.

Values

Branding iron, letter "D", wrought-iron, 21 in. long ... $14-18
Branding iron, "HL" run together as one symbol, 25 in.
 long $19-24
Branding iron, "rocking chair" symbol, 22 in. long $24-28

Broom-Making

Early brooms were made of whittled saplings, the excess wood from the thinned handle being brought down to form additional sweeping straws. The join was bound with pliable wooden strips, that wrap-all early American material. The earliest brooms, however, were simply twigs strapped to the end of a wooden handle.

Before factory-made brooms became widely available—and often, like woven-cane chair seats, they were well made by blind persons—broom-making was a specialized task. Farmers grew broom-corn, cut and dried it, and made brooms in quantity. Sold or bartered, brooms often supplemented rural income.

Values

Broom-vise, used to compress and form sweeping head,
    screw-down wooden threads, 19 in. high $45-55

Broom-cutter, for shaping broom sweeping surface,
    wrought-iron drop blade $30-40

Broom-maker, for fastening broom to handle and wrap-
    ping broomstraws, hand and foot operated $110-135

Bucksaws
(See also Saws)

Other than the general-purpose carpenter saw, and perhaps the tree-felling crosscut saw, no other farm saw was as common or more useful. And it is the one saw that has far more wood than metal, with most examples put together as if by a cabinet-maker. The word itself comes from the buck or sawhorse with which it was often associated, "bucking" firewood.

Most bucksaws have an "H"-shaped form comprised of wooden parts. The bottom of the "H" is closed with the relatively thin long blade, the top by the metal or thong tension-unit. This was shortened—to make the blade tight—in one of three ways. For threaded metal, there may be a central turnbuckle, or, a wingnut outside the handle or opposite wood frame member.

Early bucksaws and some later ones had a thong or wire tension that was tightened with a short wood winding slat. It was then permanently braced against the center cross-bar, unable to turn. The handle side of the "H" usually has a graceful curve in a mild "S" configuration, and the whole saw was generally very well made. The design was so successful that metal versions are still made, though the frame now has a simpler shallow "U"-shape.

Values

Bucksaw, turnbuckle tightener, maker-marked frame .. $15-19
Bucksaw, twin-wire tension with wooden brace, 29 in. long $17-22
Bucksaw, curved pipe metal frame, rubber hand-grip .. $8-11
Bucksaw, salesman's sample, 6½ in. long $45-55

Bullet Moulds

Undramatic, often ignored in antiques shops and at sales, bullet moulds are a proud but sometimes overlooked part of the American farming past. Once required equipment for every muzzle-loading rifle and pistol, surviving examples are becoming scarce.

Used from the 1500's into the early 1900's, most of the very early bullet moulds have either disappeared or now repose in museums. Examples, from the late 1700's through the late 1800's can occasionally be found at reasonable prices.

A bullet mould is a small metal-casting device. The actual mould cavity is of slightly less diameter than that of the gun bore for which the balls were being made. The typical mould consists of five important parts, these being the chamber, channel, handle, body and shears.

Made of hand-wrought iron or, later, of factory-cast iron, the typical mould had an almost perfectly round casting chamber. this was bored out by a steel sphere called, appropriately, a "cherry" bit or spinning head.

The channel was a small tunnel that lead from the top side of the closed mould into the chamber, funneling in the molten metal. Each ball then had a miniature channel projection called a "sprue". The handle extensions served two simple functions, making the mould easy to maneuver.

Handles also served to close the mould for pouring, and opened it for ejecting the solid ball. The body was the entire solid portion, two nearly identical halves, in the form of a blunt-nosed pliers. The halves were held together with a heavy rivet.

The shears, located close to the rivet on the handle side, was used to nip off the sprue base so the cast ball was more nearly spherical. Often the remnant of the sprue was further filed down so the ball would not waver in flight.

Pouring bullets was a careful and time-consuming operation. Bulk lead was first purchased, and an appropriate amount was melted in a ladle over a fire. This had a pouring spout, which was directed into the mould channel, and only enough lead was poured to form the ball. While not requiring any great skill, it took a steady hand to avoid lead spatters.

Lead was once at a premium, and was never wasted. The writer was informed by an old man that when he was a boy, they often used their half-stock caplock rifles for target practice. They put a mark on a sawed-off log and sharpened their shooting skills against it. When the log was riddled, it was burned in the fireplace so the lead could be salvaged and recast.

Other than the typical farm mould likely to be found today, there were others. For economy and because most moulds were used at irregular intervals, most are single-cavity. That is, bullets were made one at a time to replenish a low supply.

Some moulds have several bullet-casting cavities of different sizes. the gun owner could thus order a single mould for several firearms, rifle, pistol or shotgun. A small cavity (just less than, say, .25 to .50 calibre, or ¼ inch to ½ inch in diameter) usually means a rifle, with pistol dimensions similar or smaller.

A large cavity meant that a civilian bullet mould was used for making shotgun balls, for which an archaic system of measurement has always been used. While a calibre figure is that number over 100, giving the diameter in inch fractions, the smoothbore shotgun was different.

Such bullet moulds came in "gauges", with the size based on weight, not diameter. So, a cavity for a single-ball shotgun load might be 28-gauge, meaning the finished balls weighed 28 to the pound. A 12-gauge ball weighed 12 to the pound, and so on. To confuse this simple explanation, the more recent .410 shotgun (the .22 rifle of the shotgun family) is classified by calibre and has a base diameter of 41/100 inch.

Another type of bullet mould to be found occasionally among rural antiques and collectibles is the multiple-cavity military mould. It may have certain characteristics, like a unit mark. The body is long and heavy-duty, and there may be long wooden handles.

Usually the cavities are the same size because of uniformity in military arms, with a cavity diameter of half an inch or more. Many military moulds were made of brass; though expensive, the moulds remained free of rust and pitting. Since the Revolutionary War and until after the Korean War, surplus military arms and accoutrements have been sold to civilians, and some ended up on farms.

A bullet mould can be assigned an age period, generally, given certain considerations. Round cavities were in use until the early 1900's. Conical cavities became widely accepted about the time of the Civil War when it was learned that an elongated, pointed-nose bullet was far more effective than the traditional ball. Sometimes a single mould will contain both types of mould shape as the owner may have wanted a selection. Such moulds may date from the late 1800's.

The average bullet mould was made of iron, but many recent examples are iron or steel, plus some of cast brass. Some very different moulds were actually hand-carved from steatite or soapstone, and these tend to be both old and rare.

A soapstone mould is usually rectangular, and almost always there are several cavities for round shot. Each half of the mould bear similar impressions, as, half a ball cavity, half a pouring channel. Originally, and in use, the halves were peg-fitted together for cavity alignment and the whole tightly bound with rawhide straps. Most examples turn up in Eastern parts of the country.

In early days, making bullets in quantity was such a laborious task that a few persons were able to earn extra income by taking over such work. There is a surviving 1775 account, from Pennsylvania, of one man who advertised "Musket Balls". Usually, however, bullets were made by the user, and this was just one more process that rural Americans had to master.

Bullet moulds were once common farmstead objects. Metallic, self-contained cartridges became widely accepted in the late 1800's, and moulds were thrown away in numbers. They are still being found in outbuildings and in farm dumps.

So far, fraudulent bullet moulds are not much of a problem.

But other moulds—those made for the "blackpowder boom" beginning in the 1950's—can sometimes be offered for early types. Made for service and not really as reproductions, the metal will generally be in bright condition. Even though the moulds have the old, basic styles, they still look, and are, recent.

Values

Bullet mould, wrought iron, about .35 calibre, rust-pitted	$10-13
Bullet mould, wrought iron, fine condition, 8 in. long	$15-25
Bullet mould, soapstone, very early and rare, three cavities	$75-90
Bullet mould, about .40 calibre, wood handles, 9 in. long	$16-25
Bullet mould, cast iron, fitted sprue-cutter lever, 5 in. long	$15-22.5
Bullet mould, marked "Winchester", sprue-cutter, walnut handles, brass sleeves, orig. cavity filled and rebored	$30-40
Bullet mould, all cast brass, two different sized casting chambers, 4½ in. long	$35-45
Bullet mould, "Ideal", detachable casting block, cal. .467	$20-27.5
Bullet mould, for Remington pocket pistol, iron	$35-42.5
Ladle, lead-casting, cast iron, covered spout	$15-20
Ladle, lead-casting, wrought iron, wooden handle	$30-35

Burl

Burl is the gnarled mass that protrudes from the trunks of some hardwood trees. The wood is very dense, and the grain twisted and convoluted, making anything worked from it both attractive and extremely tough and durable.

Values

Burl bucket, one piece of wood, 6 in. high	$65-80
Burl bowl, good grain and finish, 16½ in. diameter	$450-500
Burl bowl, ash, 12 in. diameter	$375-400

Burl bowl, ash, two handles, deep, superb condition,
6¾ x 16 in. diameter $1025-1200

Burl bowl, footed, 3⅛ in. high...................... $225-250

Burl bowl, turned, 2¼ x 3⅝ inches.................. $135-150

Burl dish, ash, worn finish, 4 in. diameter........... $305-350

Burl funnel or jar filler, with wear and cracks, 3x5
inches $25-35

Burl mortar and wood pestle, 6½ in. high........... $100-125

Burl scoop, primitive, 9 in. long $35-40

Burl stomper with hickory handle, 29½ in. long $55-65

Butchering Equipment

Fall butchering time was an important and almost festive occasion on the farm, despite the fast pace and hard work. A large amount of meat had to be prepared in a relatively short time, some cooked, some made ready for curing and smoking, more to be mixed and seasoned for sausages.

Additional cooking and treatment produced delicacies like scrapple, made of meat scraps with added corn meal, generally sliced and fried. Head cheese was made, a jellied mass of chopped and boiled hog parts, as was meat "puddin' ", of minced and cooked meat.

Butchering required far more utensils and implements, large and small, than is generally known today. Most farmers kept the main implements since they were needed every year. Some thrifty souls borrowed at "butcher-time", since the tools were only needed once a year.

Large containers included scalding troughs up to six feet long, usually made of planks. Boiling water was added as needed, this coming from the kettles. These were made of cast-iron, copper, even brass, with the last the smallest. Older children had an all-day job in bringing water for the kettles, and fresh wood for the open fires under them. Whatever the type, kettles were either hung in a line from a heavy beam, slung from a tall, wooden tripod, or supported on an iron stand.

A wide array of knives was used, from the standard butcher knives with straight edges and many lengths to curved-edge skinning knives and the thin-bladed boning knives. A unique, long sticking knife was used to kill hogs and beef cattle, and it had a duo-edged triangular tip.

Meat saws of several kinds were used, mostly for cutting bones. The cleaver-wielder was often the most experienced man as it required both skill and stamina to be used without wasting meat. Cleavers from one to five pounds were used, either on chopping blocks or the large butcher-block surface. To keep edged tools razor-keen, whetstones, sharpening steels and large grindstones were in frequent use.

Meat-hooks of a number of kinds were used, from suspending whole carcasses to dragging and carrying meat slabs. Most meat-hooks, while sturdy, have small curved ends and are quite different from bale-hooks. Hooks include those with wood handles for one hand, two hands, and, extra-long for two men.

Another class of hooks was used in the smokehouse for hanging meat in the aromatic corncob or hickory fumes. Multi-pronged types hung from rafters or ceiling, and some single hooks were beam mounted. A "meat tree' might arise from the floor or dangle freely. First of wood, they were later made of wrought-iron.

Hog scrapers, simple iron disks with a central wood handle were everywhere, as were the "gambrels," or hog-spreaders. These were the wooden bars from which hogs hung head-down. Short hardwood stakes were sometimes placed to open the carcasses and aid in rapid cooling.

Meat grinders or choppers were absolutely necessary, grinding both prime cuts and odd bits for the creation of many meat dishes. Such grinders began with wooden cylinder types and hand-forged blades, with meat chunks going in the top of one end and emerging, ground, from the bottom of the other end. Folklore also suggests that such choppers were also used to dice pickles. These handmade choppers are both early, ca. 1850, and eagerly sought.

Later came the many grinders, factory-made, that resemble the common up-sized kitchen meat-grinder. Some butchering-type grinders could handle 300 pounds in an hour, all via the simple crank handle. Various holed wheels could be inserted to control size of the grind. And some had a tubular extension so they could also be used to stuff sausage.

Sausage-making however, was usually done with a separate machine with a simple purpose. It compressed the meat mix in a hopper and forced it through a spout into a sausage casing. It could emerge as long single links, or be hand-twisted to form shorter sections, all to be eaten fresh-cooked or smoked. Seasoning, in the form of herbs and spices was often added, though purists insisted only on a dash of salt and red or black pepper.

Sausage-stuffers include the tubular direct-plunger type which provided a fairly loose fill. For a tightly packed link, lever and screw operated stuffers created more pressure. Lever-type stuffers were usually plank-mounted, and an arm pushed the plate through the hopper.

The most efficient stuffer was probably the low, cylindrical kind with the pressure plate powered directly by a top wheel or indirectly by a geared side crank. These had good capacity and control, with the spout emerging from the side-bottom. The spouts alone are collectible, being tin, chromed sheet iron and even copper or brass.

Many other butchering implements can be found. They include the unusual puddin' workers, wood-handled, iron-shafted with right-angle turns and various blades at the end. They were used to stir and scrape the interior of iron kettles so the meat mixture would not settle or scorch.

Hand-choppers, skimmers, tasters, pickling vats, smoke-flavor injecting syringes, casing trays for cleaning hog entrails, strainers—all were part of the madness and magic of butchering time.

Values

Knife, sticking, 7¼ in. blade, hardwood handle $11-15

Knife, skinning, 6½ in. blade, brass-riveted handle.... $10-12

Knife, butchering, 13 in. blade, curved edge $12-15

Sharpening steel, rusted, wood handle, 14 in. handle .. $2-4

Saw, meat, wood handle, steel frame, old blade $5-8

Cleaver, cast-steel blade, hardwood handle, 13 in. long $15-20

Chopping block, laminated maple, 15x24 inches $25-32

Butcher block, iron-strapped hardwood, 26½ in. high . $110-140

Meat-hook, one-hand, hardwood handle, 8 in. long.... $7-10

Meat-hook, two handed, 10 in. long................. $9-12

Hook, for hanging, wrought-iron, eyed, 7 in. high $20-28

Hook, for hanging meat in smokehouse, single prong,
    5 in. high $8-11

Meat tree, wood, smokehouse, floor-mounted, eight
    projecting wood arms, 7 ft. high $65-85

Meat tree, ceiling hung, wrought-iron, with suspension
    hooks, 2½ ft. high $80-110

Meat grinder/chopper, primitive, round wood casing,
    wood handle, iron blades, 23 in. long $110-135

Meat-grinder, plank-mounted, cast iron, 15 in. high . . . $25-35

Sausage-stuffer, tin cylinder body and spout, wood
    plunger and cap, wood handle $32-38

Sausage-stuffer, wood, iron and tin hardware, 47 in.
    long $30-40

Gambrel, made from crooked tree limb, whittled $3-5

Gambrel, hickory, three end-notch positions, metal clip
    at center for hanging $7-10

Gambrel, walnut, two end-notch positions, well-made . $9-13

Hog scraper, iron, wood handles, 5 in. high $12.5-15

Hog scraper, tinned iron, rusted, wood handle $6-9

Utensils, five-piece set, wrought-iron, scrolled handles;
    small and large dippers, spatula, strainer, and
    fork $275-300

Hand-chopper, single iron blade, rounded edge, metal
    handle $11-15

Hand-chopper, double parallel blades, walnut handle . $23-39

Casing tray, solid cherry wood, 4 ft. long $135-175

Butter-Related

Very many butter-related collectibles exist because butter-making was big business on the farms of yesteryear. In 1849, farms produced over 313 million pounds of butter, which increased to 777 million pounds in 1879. In 1918, established central creameries first outstripped rural production, 823 to 710 million pounds, respectively.

Overall, a great number of such tools and implements can be tentatively dated in the 1800's into about the 1940's, with some specimens much earlier. Butter, the end result of agitating slightly-soured cream, has a nearly endless range of associated objects.

Bowls include the familiar turned round wooden examples, perhaps 8 to 20 inches in diameter. Some had a copper strip tacked around the outer rim to retard splitting. Oblong examples were often hand-hewn, with the bowls being made in many styles. Butter bowls include large ceramic types, some of which were made in nesting sets. Often such "mixing" bowls were used to work butter, and examples in spatterware or spongeware are highly sought-after.

Butter carriers are not common, and may have originated in Pennsylvania Dutch country. Of wood, often pine, many were handled and had space for half a dozen or more one-pound butter containers. These can be considered market carriers.

Open-topped stoneware crocks were used for both fresh milk and sweet and sour cream, as well as for butter storage. The top would be covered with waxed paper or a common table plate to protect the contents.

The keeler or "keel" was a small wooden tub, and the name may have come from "cooler", which was their function. These stored milk (often in the springhouse or milkhouse) until the cream rose and could be skimmed and collected.

The term ladle or even scoop is applied to many butter implements that obviously also served other functions. Avoiding misunderstandings, the ladle is a wooden object with lower blade and handle, likely used to transfer butter from one place to another.

Butter molds or presses served three functions. Since butter normally is simply a yellow mass, the mold served to compact, even decorate, the butter. Most molds (except miniature examples) also helped in marketing, for they often pressed exactly one pound. The molds exist in wood mainly, but glass and ceramic examples can be found, also a few in a heavy grey metal.

Molds tended to be either round or rectangular; they consist of a hollow body to hold butter and a movable plunger to compact and/or mark the top. The plunger face often had a carved-in design with a simple motif like wheat sheaves, stars or flowers. Scarce examples may have animals or fish. Some rectangular molds had hinged sides to aid in releasing the pressed butter.

Large butter bowl about 16 in. across, with unusually thin sides for size, turned wood, excellent condition. $45-60

Butter paddle, unusual example with both holed and initialed handle end, about 10 in. high. $27.5-35

Butter mold, round one-pound size, with star-marked plunger face, carved softwood. $40-50

Butter mould, one-pound rectangular size, factory-made with dovetailed corners, hardwood. Owner's initials are on the pressure plate; mould is marked with original paper, "Selected Michigan Hardwood/The Munising Woodenware Company". $40-45

Butter mold, extremely fine condition, rectangular, dovetailed sides, floral motif on plunger. $55-70

Butter-buyer's sign, with prices on fabric tape for changes. Heavy cardboard; if rate were adjusted to probably age of sign, the value of butter per pound would be closer to sixteen cents. $16-20

Photo courtesy The Peddler Antiques, and, Thomas Collection.

Hand-carved softwood butter paddle, 9 in. long. $6-9

One of pair of "Scotch hands", for making butterballs. These could also be used as paddles and scoops. Matched pair $16-22

Butter scoop, slightly concave wide blade, edged, handle-end hook for butter bowl, about 8 in. long. $14-19

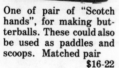

Oak handcarved butter spade, well-used. $11-15

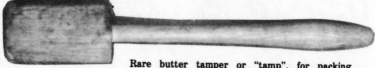

Rare butter tamper or "tamp", for packing shipping containers; octagonal, heavy head, hardwood (prob. maple), from Ohio, 15 in. long. $40-50

(To beware of in this area are plain-faced plungers that have had a recent design carved in, and plungers that are too large or small for the original mold sides. Unfortunately, considerable alterations and mismatching exist, as do reproductions.)

Paddles tend to have a wide, thin blade, straight or slightly curved, and the handle end may be hooked for the side of the butter bowl or for hanging.

Butter prints or stamps often had a turned, knobbed handle, and were used without a mold. They could be applied "freehand" to any butter mass to produce a full or partial intaglio design. Duo-handled rollers are related, and resemble a small rolling pin. One or many designs were cut out in the central cylinder. Used to imprint a flat sheet of butter, the decorated sections were later cut into smaller servings. Such work was often done in the out-building called the summer-kitchen, always near the back door of the farmhouse.

Butter-weighing devices are known, many tending to be both old and wooden. Such scales were of the balance type, and had suspended platforms and a central shaft. Butter skimmers of horn, wood or tin were used to remove cream or take out fallen debris.

The so-called "Scotch hands" were made in matched pairs. These resemble paddles, except that one flat surface has tiny longtitudinal grooves and ridges. These were used, corrugated sides together, to produce butter spheres, or butterballs.

The butter smoother, rather scarce, was used to shape sides and top of a butter mound. They gave a uniform appearance to market butter, and one country store owner in the late 1800's is known to have used one to erase telltale mouse tracks. Of wood, some were 4 to 6 inches long.

Butter spades do resemble the garden spade, having a long, narrow, straight blade and straight handle. They were used for removing deep, narrow butter sections, and the rounded working edge was helpful in scraping containers.

The tamper or "tamp" resembles a kitchen masher, but with a handle up to 15 in. long. Made of hardwood, sometimes maple, the head may average 3 in. in diameter. These were used to pack container butter. Often such long-handled curiosities can be discovered by simply sniffing the bottom contact surface. The years of pressing butter leaves a faint but distinct aroma.

A butter tester was a thin metal device 16 to 20 in. long, designed to be twisted so as to remove a butter core. This was a quality check or safeguard, so that old or rancid butter was not layered under fresh.

Similarly, long probes of hickory or metal were also used to be certain that the butter did not contain stones or other foreign objects. In the days of barter or cash sales, one had to be careful not to get "deaconed", or cheated. Butter was once a very common item of farm trade, and widely used.

Tubs were made by some manufacturers in up to eight sizes, and were often of softwood like cedar. Most were stave-constructed and iron-strapped. Butter-working tables were special-made to knead butter, removing air bubbles and excess water, the true buttermilk. A number had corrugated working tops and a matching roller with side crank handle.

Butter churns exist in countless styles, but basic types are listed. There is the round or square glass-bodied table churn, with metal top and circular toothed gear, wood handle and wood paddles or dashers.

A much earlier "plunger" churn was a floor model, circular, large on the bottom and small at the top. A round wooden plate covered the top, and a hole at center permitted the up-and-down movement of the pole-operated agitator. Stave-constructed, earlier similar churns had wooden hoops while later types had iron hoops.

Larger "rocking" floor churns with four legs can be found, and the barrel body was turned on an axle through the center. Inside, angled projections helped make the butter "come". An unusual type of floor churn could be the product of a back-woods inventor. This was a plunger churn in a frame, with the agitator turned with a wound cord attached to a lever. The butter was thus worked by a swirling movement.

A little-known aspect of butter-related items are the metal extenders or recombiners. These employed left-over butter and milk to quickly produce a fast-butter that was acceptable for table use.

One example, called the "Miracle Merger" used one-half pound of creamery butter and one-half pound milk. Multi-purpose, the tin cylinder with plunger also whipped eggs or cream.

A larger version, with about three times the capacity of the merger, greatly resembles a crank-type home ice-cream maker. The principle is the same, the mechanical process different. Most such devices seem to be factory-made, from the early 1900's, and more interesting than practical.

Values

Butter bowl, turned wood, 11 in. diameter	$13-18
Butter bowl, turned wood, 15 in. diameter, hardwood burl	$45-75
Butter bowl, spongeware, 11 in. top diameter	$55-80
Carrier, pine wood, for ½ doz. 1-lb. blocks	$70-90
Churn, square glass body, metal working gear	$30-40
Churn, plunger-type, metal-banded, factory-made	$75-100
Churn, plunger, wood-strip bands, early	$90-125
Churn, plunger, set in frame, thong-turned handle	$115-150
Churn, barrel-type, stave-cosntructed, fine condition . .	$100-150
Churn, floor-rocker type, 3-5 gal. capacity	$125-175
Crock, pale white stoneware, unmarked, 11 in. high . . .	$18-25
Extender/recombiner, "Miracle Merger", 1-lb. size, tin	$30-40
Extender/recombiner, cylinder, crank, metal	$35-45
Keeler, staved, 5 in. high, 13 in. diameter	$30-45
Ladle, plain, wood, handmade, 9 in. long.	$12-20
Ladle, factory-made, wood, hooked handle end	$15-25
Mold/press, rectangular, plunger face with 3-cherry design	$30-45
Mold/press, round, 1-lb. size, face with wheat design . .	$40-60
Paddle, wood, 9 in. long, 3 in. wide blade.	$8-12
Print/stamp, flower design, round, 4 in. high	$45-65
Print/stamp, on opposite side of wide-blade paddle, softwood, star design, 10 in. long, rare	$125-175
Probe, hardwood, rounded end, thicker handle, 22 in. long	$8-15
Roller, 4 different designs, wood cylinder, 14 in. long . .	$75-95
Scale, wood, balance-arm type, old, 20 in. high	$80-125
Skimmer, tin, oblong, 5 in. average diameter, punched-out design	$9-12
Scotch hands, 10 in. high, well-used, soft wood, pair . . .	$14-18

Scotch hands, 13½ in. high, hardwood, maple, pair ... $25-35

Smoother, hardwood, rectangular, rounded surface,
    7 in. long $15-25

Spade, 11 in. high, blade 6 in. long, 1¾ in. wide $8-12

Table, factory-made, wooden cogged cylinder, about
    18x26 in. $80-150

Tamper, maple, 6-sided head, 15 in. high $35-45

Tester, metal, 17 in. long, scarce $45-75

Calf Weaners

To prevent calves from suckling when the farmer thought it was time for other foods, some contraptions appeared that helped solve the problem. In most cases, field grass or hay could still be eaten through the weaners. The simplest type was basket-like, and was fastened to the calf's muzzle. It raised to permit grazing, but dropped to prevent suckling.

A more dangerous type of weaner was designed to make the mother reject her offspring. Rather than teach the calf to forage for itself, this was an iron and wire mask with a number of projecting nail-like spikes.

When the calf attempted to nurse, the cow responded with rapid movements elsewhere, or kicked the calf away. This weaner was controversial as some cows were injured by hungry calves. Most farmers eventually arrived at the best solution, separate penning facilities.

Values

Weaner, splint-oak, early, basket-type, damaged
    condition $13-18

Weaner, basket-type, galvanized metal, with leather
    head-strap $8-13

Weaner, frame-and-prong-type, tinned metal $7-9

Weaner bucket, surrogate-teat, rubber, on 2½ gal.
    capacity container, 14 in. high $8-10

Calipers

A caliper (or calipers) is used to note interior or exterior measurements, which it does with two curved metal legs. They exist in wood or metal, and were useful for tasks like making and fitting a wagon axle to the wheel hub. Collectors look for marked examples, those from select or little-known makers, and the presence of brass.

Values

Calipers, leg, "Starrett-Jenny", 6 in. long	$8-10
Calipers, double, iron, 18 in. long	$70-90
Calipers, wrought iron, one side with measure-marks, 20 in. long	$50-65
Calipers, "P.O. Lowentraut Mfg. Co./Newark", 4 in. long	$8-10
Calipers, "Starrett/Pat. 4-16-1901", 4½ in. long	$8-10
Calipers, "Union Tool Co./Orange, Mass", 6½ in. long	$10-12
Calipers, "Stanley/#36½", brass and boxwood, brass hinge	$25-30

Carts—Farm
(See also Animal Carts)

People-powered carts were often a farm feature after 1900, and were used to propel relatively light loads for short distances. Milk-carts were of several styles, and the two-wheeled type conveyed full cans of milk or cream. It had a very low, flat platform, loaded and unloaded from the front, and was pushed and steered with the horizontal connected handle bar.

The milk-barrow resembled a wheelbarrow, except that it too had a very low but narrow platform. Due to the single front wheel, this push-cart loaded from the rear and was guided with the two projecting wooden handles.

The frame cart was a metal skeleton consisting of axle, tongue and connecting handle, with enough interior space to fit a barrel. This in turn held anything from orchard spray to "slops" or feed for hogs. Such push-carts were generally used only between or near the outbuildings.

The box-cart or general-purpose cart resembled an early city push-cart, and could be used for garden produce. The bed was just above the high axle, and it was a simple wood-frame box with sides. (It's larger equivalent was the horse-drawn farm wagon.) As with most farm carts, the wheels were metal, about 3 ft. in diameter.

Wheelbarrows generally had the shape we know today, but early types were all-wood, including the front wheel with turned spokes. Often there were two on the farm, one for heavy work and a smaller, lighter wheelbarrow that could be used by women and children for yard and garden work. And often enough, to the consternation of youngsters, their play pull-wagons became, in an emergency, a working vehicle to be loaded down.

Values

Milk cart, all-metal frame, wheels and platform $25-32

Milk-barrow, twin wood handles, short metal plat-
form $30-40

Box-cart, joined handle, wooden side extenders $50-60

Wheelbarrow, lightweight, late-1800's, painted line
trim, fine condition $100-125

Cattle Leaders

Taking cattle from one place to another on the farm required both knowledge and equipment. Older, well-trained animals could be guided by voice or light taps from a cane, and farm boys sometimes became very accurate with a slingshot. But often, cattle simply would not go in the right direction.

So, a variety of cattle leading or directing devices developed, from primitive to electric. The neck-lead or cattle-tie was favored and began as a leather or rope loop. Later a manufactured chain with large catch and elongated holding link was produced, with a trailing chain for leading or fastening. These were handy for taking cows to a new pasture or tie-downs for a vet inspection.

For gentle leading, a nose-lead was used, this a tong-like scissors-action tool that closed when the brass shank-spring was released. Most factory-made types are quite similar, and they are rarely maker-marked due to the small size and rounded surfaces.

A more potent instrument was the circular bull-ring, 2½ to 3 inches in diameter, either of steel or cast and polished copper. The rings were hinged and opened with the free ends sharply tapered to points. For attachment, the ends were driven through the soft nose septum of the bull, then permanently closed with a small brass screw.

The bull ring in turn was held with a heavy snap connected to several feet of chain or heavy rope. This terminated in either a hand-loop or a short wooden handle so the bull could easily be led without danger of being gored by horns.

Cattle-prods were of several types, beginning with light whips and pointed sticks. Unrestrained animals could then be guided through gates or onto the chutes of loading docks. In the early 1940's a new device, the electric prod, began to be used on some farms.

One common type was little more than a giant flashlight with a 1½-ft. metal projection, terminating in a "Y"-shaped contact point. Harmless, the push of a button delivered a high-voltage jolt to sensitive areas, and a stubborn animal usually responded by moving in the right direction.

Values

Cattle-tie, linked, iron, elongated chain, oval tie-link ..	$4-6
Nose-lead, iron, brass wire closure spring	$2-4
Bull-ring, 3 in. diameter, polished solid copper	$5-8
Bull-ring, steel, about 2¾ in. diameter, with 2 feet of heavy chain..................................	$6-9
Cattle-prod, electric, used five large-size flashlight batteries, corroded case, ca. 1944	$7-10

Cheese-Related

Some farmers specialized in making cheeses, and the associated collectibles are unique. Most were handmade, even into later times, with non-contaminating wood the favorite material. Testers and knives were often factory products; the rural cheese-making tradition often followed that of the Germans and Swiss.

Values

Cheese basket-strainer, woven open-work, 6½ in.
    diameter $145-160

Cheese basket, for market cheese, splint, 25½ in.
    diameter $120-135

Cheese-box, bentwood, for small wheel, lidded, 8 in.
    high $30-40

Cheese cutter, pole with spaced wire cutters for stirring,
    metal, 41 in. long $35-42

Cheese dryer, woven splint, bentwood frame, 22½ in.
    diameter $70-80

Cheese knife, triangular blade, wood handle, 9 in.
    long $12-16

Cheese ladder, old, wood-pegged and mortised frame .. $35-40

Cheese press, wood, for circular wheels, bentwood com-
    partment, 20 in. wide $70-80

Cheese press, wood, shoe feet, pegged construction,
    21 in. wide $80-95

Cheese tester, wood twist handle, tapered iron blade, for
    removing sample cores, 15 in. long $35-45

Chicks and Chickens

Almost all farms of any size and acreage had two distinct out-buildings devoted to poultry culture. One was termed the brooder-house, the other the hen-or chicken-house. The first raised the hatchlings, chicks or "peeps", while the second was the standard chicken-house with feed and water facilities, a roost, and nest boxes.

Before such refinements, the production of baby chickens was entrusted to a few brooder hens, each with a clutch of 12 to 16 eggs. To lower the attrition rate, the farmwife or children might bring the chicks into the house and resettle them under the range or coal stove.

The brooder house might have a 50- to 200-egg incubator, perhaps by Imperial or Wyandotte, warmed by kerosene and hot water or air. Fine objects of good wood and copper, a few still turn up on the market today; incubators in good condition are far from common.

Large wooden cheese press, Kentucky, used to press whey from cheese. Note the bentwood cheese form or box near spout end.

Photo courtesy Museum of Appalachia, Norris, Tenn.

Right, Chicken waterer, one-gal. stoneware container, side opening, white, glazed, 14 in. high.

$18-24

Hen and chick shelter; the hen was prevented from digging in gardens while the "peeps" could hunt food and return for protection.

Photo courtesy Museum of Appalachia, Norris, Tenn.

Cattle equipment; top, fastening chain, for around neck and ground tie, factory-made. $5-7

Cattle nose-lead, cast-iron, brass spring for closing, lead-rope missing. $3-4

Above
Clamp, all-wood, large size with 14 in. jaw lengths, wood-thread worms. From New England, and with owner's initials burned in sides.

$12-15

Right
Iron corn-row stake set, used with trip wire to space seeding with automatic horse-drawn corn seeder. The pair ... $14-19

Farm grinder, "Model No. 2", for cracking corn, grinding charcoal, powdering oystershell, etc.
$15-19

Chisels, unmarked, top specimen 10 in. long.

Top	$2-4
Center	$1.5-3
Bottom	$4-6

Chick waterer; one-qt. jar set on zinc base. $5-7

Feeding apparatus includes galvanized metal trays and hoppers, scaled down to have openings just above the floor. Waterers, or "chicken fountains" include screw-on types for canning jars and factory designs in both metal and pottery.

When chicks could fend for themselves, they were moved to the larger chicken-house, which opens another realm of collectibles. Bigger feed troughs were needed, some very well made of scrap farm wood. Flaked oyster shell was fed for egg-shell calcium, and there were containers for this.

For marketing, shipping flats or cages with dowel bars were on every farm, and could hold about a dozen birds comfortably. For capturing individual chickens, a device called the chicken-catcher was employed. This was a long, heavy wire with a narrow crook-like hook at one end. It was used to catch and hold the nearest leg of an unwary fowl. Chicken-catchers exist in both hand and pole-mounted sizes, and had no other use on the farm.

Blunt spur-caps were sometimes placed on belligerent roosters to protect other chickens, and a few were made of handcarved wood, secured by thongs. Perhaps surprisingly, as farm outbuildings are demolished, there seems to be a market for wood nesting boxes, individually and by the row.

Values

Chick fountain, pottery, black glaze, 8¼ in. high $20-25

Chicken fountain, stoneware, blue chicken transfer
    design, 9¾ in. high $25-30

Poultry-killing tool, cast brass, steel blade, factory-
    made, 6½ in. long $35-45

Incubator, wood cabinet on legs, 39 in. high $65-80

Chick fountain, zinc screw-on bottle base, 6 in.
    diameter $2-3

Chick feeder, galvanized metal, feed slot on one side,
    19 in. long $9-12

Feed trough, refinished, perch bars, wood, 15x18x41
    inches $25-35

Oyster shell container, wood, cut from standard
    barrel bottom, staved, homemade $12-15

Shipping cage or flat, good, clean condition $20-25

Chicken-catcher, wire, 19 in. long $2-3

Chicken-catcher, wire head, 3½ ft. pole handle $5-7.5

Spur-caps, for rooster, wood, without fasteners, early . . $6-9

Nesting boxes, set of 12 on three levels, four wide,
wood, clean and well-made $70-90

Nesting box, single unit, 14½ in. high, wood $8-11

Chisels

Square-edged chisels have a straight cutting edge, small in proportion to overall length. All metal, of iron or steel with wood handles, they could either be used with hand pressure or be struck by a hammer or mallet.

Multi-purpose, chisels could cut the rectangular holes for wood building frames or make the openings for flush-set door hinges. Chisel-work was needed for removing excess wood from small portions of hard-to-reach places, and every farm had an assortment.

Values

Chisel, ³⁄₁₆ in. blade, "Fulton Special", 12½ in. long. . . . $16-18

Chisel, swan-neck, brass ferrule, ⅜ in. blade width $24-27

Chisel, corner, "O.V.B.", 15½ in. long $27-30

Chisel ¼ in. blade width, "Lakeside", 9 in. long $19-24

Chisel, 1 in. blade, "Greenlee", 14 in. long $22-26

Chisel, 1¼ in. blade, "Barton", 15 in. long $25-30

Chisel, ⅞ in. blade, "Douglas Bottom & Co.", 9½ in.
long $22-28

Chisel, 1¼ in. blade, 'P. S. & W. Co." $27-32

Clamps

Clamps were invaluable for making, repairing or holding other objects. The first clamps were rough wood, jaws held by a tightened rope or thong. Later came all-wood examples in many sizes, with jaws that opened and closed by turning large wooden screws. At first the threads were handcut, this done until a screw-cutting tool was developed.

Metal clamps of a wide range of styles were found on the farm, ranging from hand-held to types that fit into holes at the workbench or at the anvil. Most clamps, excepting those of trades like leather-working, could hold a wide variety of objects.

Values

Wood, two rectangular jaws, turnscrews, 14x19 in., owner's initials burned in $25-30

Wood, with wood turnscrews, 11x13 inches $20-25

Wood, with metal turnscrews, 5x8 inches $7-10

Metal, japanned wood handle in black, iron turnscrew . $9-12

Metal, clamp base, jaws 7 in. long, multipurpose $3-4

Metal, insert base, 5 in. jaws, faint maker's mark $2-3

Metal, "Jint Co.", 10 inches, lever-style tightener $18-24

Metal, "E. C. Stearns & Co./#4/Syracuse, N.Y." $28-34

Clapboard Tools

Many frame farm buildings were covered with clapboard when lumber mills became numerous. Clapboard ran parallel with eaves and foundation, with upper edge thinner than the lower. This in turn overlapped the clapboard below. Clapboard siding was weathertight and simple to put up, providing one had several tools.

The clapboard marker adjusted to the width of the board and had teeth that could be set to mark the board for sawing to fit properly at the corner of the building. The clapboard gauge aided in setting each board (from bottom to top) and permitted the same amount of overlap each time. Both types were at first hand-made, followed by factory production.

Values

Clapboard marker, metal body about 6 in. wide, factory-made $14-19

Clapboard gauge, adjustable, metal underlips, thumb-screw $16-20

Clapboard gauge, rigid, hardwood, for laying up two boards at a time, 13 in. high $35-45

Conestoga Wagons

(The writer is indebted in large part to Ivan Glick, of SPERRY/NEW HOLLAND, New Holland, Pennsylvania, for basic information on the famous Conestoga wagon.)

The Conestoga was a heavy covered wagon used for both work and travel. Believed to have been first-built in the Conestoga Valley and surrounding areas of Pennsylvania, it had wide wheels with the rear set of much larger diameter than the front set.

The box was large and high; the ribs or hoops that supported the canvas top were typically higher at front and rear, giving this wagon a certain sway-backed appearance.

The unique form possibly developed in Europe (parts of southern Germany and in Switzerland) in the late 1600's. It was definitely first an European design that arrived almost fully developed in the New World.

Interestingly, the Conestoga appeared before any significant numbers of English coaches arrived here, the time being well before the Revolutionary War. As with the Pennsylvania Dutch, the driver's lead horse was to the left. The driver handled the reins or "lines" from the left side of the wagon, which he drove on the right side of the road. This established the American road-traffic pattern that is still followed.

It should be noted that the long lines of wagons that headed West in the 1800's were not Conestogas, but much-later versions of them. The original Conestogas were locally made of readily available materials in Pennsylvania.

The Conestoga was actually a multi-purpose conveyance. The box could be removed and replaced with "hay ladders"—wide, low rail frameworks with a central plank. It was used to haul hay and sheaves of grain to the barn. After haying and harvest, the box went back on the running gears. Changing was easy, since the ceiling of the wagon shed had a permanent built-in heavy-duty windlass.

The wagons also added a word to the American language. Conestoga drivers smoked a mean little cigar that people referred to as a "Conestoga". The word was eventually shortened to "stogie", and it survives today.

In the early 18th Century the Conestoga wagon became the freight and personal-transportation device of the Pennsylvania

Dutch. It had fine engineering excellence, provided durability and strength, plus it was more than adequate for the purposes for which it was built.

There are surviving Conestoga wagon examples, as at Ohio's National Road/Zane Grey Museum near Zanesville, and, on a few farms that once belonged to early settlers. Wherever found, Conestoga wagons remain rare and beautiful objects from our rural transportation past.

Values

Conestoga wagon-jack, old red paint, wood and wrought-iron $100-125

Sidebox, ornate wrought-iron furniture and cover-lid design $200-225

Axe-holder, wrought-iron, fastened to wagon, sturdy, ornate $80-100

Whip socket, for base of whip by seat, wrought-iron, 11½ in. high $40-50

Wheel chocks, pair, for sliding under wheels when going downhill to act as skid-brakes, the set $100-130

Corn Cutters
(See also *Corn-Related Collectibles*)

Corn cutters receive a special listing because of the large number that exist, reasonably priced. They range from home-made curiosities to fine, well-balanced factory wares, and they are still being used on the occasional back-country farmstead.

Also known as corn knives or corn slashers, these are basically a long, slightly curved iron or steel cutting edge, with the reverse edge less curved and blunt. Most are quite heavy so that blade momentum helped cutting corn stalks. The handle is always wood, secured by wire, screws, rivets, or rattail haft extension.

Handmade, often very homemade, examples have the blade formed from a pounded-out file, reworked scythe blade, military sword or metal of unknown origin.

The corn blade was used with one hand to cut standing stalks of field corn. The stalk was turned into fuel, insulation or fodder, and the ear was kept for the corn seed. Manufacturers in states as diverse as Maine and Wisconsin specialized in this farm tool, also useful for taking down weeds and light brush.

A rare corn cutter is the boot-blade, a short, heavy outward-facing device which was strapped to the shoe or boot and lower leg. Logical, but never widely popular, the worker could sever a stalk with one kick, leaving both hands free for maneuvering stalk and ear.

Values

Corn cutter, hand-forged blade, carved wooden handle $8-15
Corn cutter, factory product, machined handle $6-12
Boot-mounted blade, factory made, with original straps $35-50

Corn-Related
(See also Corn-Knives)

The story of corn, once the Indians' maize, is almost the diary of early rural America. Corn was a barter or cash crop, and the mainstay of the pioneer farmstead. It provided food for the cold months and allowed the keeping of livestock. No family ever went hungry while a supply of yellow ears remained in the corn crib. Farmers once had a goal of getting corn into crib or shock by Thanksgiving Day, the mark of any conscientious man.

It is not surprising that many corn-related items exist, all associated in some way with this field crop. Some, like the hand-held seed planters, are obvious. Others, like shock-rope weights, are obscure.

Corn equipment begins with seeders, after the soil has been prepared. The first device was the planting stick, simply a sharpened pole or branch to hole the ground for kernels. _ _

Rows, even for hilled corn, were kept straight by the stake-and-string method. One advanced example is a pair of thin iron bars, pointed at one end, with footrest and coiled-wire springs for line tension.

Another pair were plain bars with a twisted center for the wire or string. Though retained for hand use, such specimens, generally in pairs, originated as spacers for horse-drawn corn planters, depositing seeds at regular intervals. With such "checkerboard" corn spacing, the plants could be cultivated in four directions.

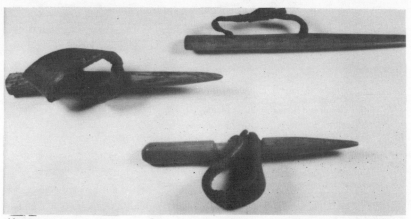

Above,
Corn-husking pegs, each handcarved hardwood with
leather finger-thongs; longest example, 5 inches. Each
$3-5

Cornbread pan, reverse side, "Griswold Crispy Corn
Stick Pan /Erie, PA, U.S.A.". $25-30

Right
Hay seeder, tin drop-spout, fabric seed bag, "Horn Self-Sower",
spout 18 in. long. $15-20

Photo courtesy Fairfield Antiques, Lancaster, Ohio

Corn-related collectible;
small popper, sheet-iron
box, wire handle, "Thrift".
$10-14

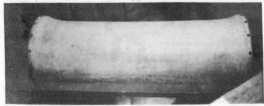

Above
Cream-related, wooden advertising measuring stick; "Buyer of
Sweet and Sour Cream", 6 in. long. $3-5

Below, Canvas bag set inside flat wooden platform. With
a central neck with tie, this is a chicken delouser; the bird
was put inside with disinfectant powder and the whole
was turned with a side crank.
Photo courtesy Museum of Appalachia, Norris, Tennessee

Right
Cream pail, tin, lidded, wire-bail handle with japanned
wooden handle.

                                                    $12-15

         Photo courtesy Fairfield Antiques, Lancaster, Ohio

Below
The old Sharp corn mill; note milling stones and wooden shingles, all original.

            Photo courtesy Museum of Appalachia, Norris, Tennessee

Other factory-made seeders were used, mostly wooden with twin handles and a metal "V"-shaped hole-maker. Pushing the handles together opened the bottom and several kernels were deposited. These were used to plant small parcels of land, like the garden or "truck patch". They were also employed to reseed any skipped areas in rows after the drawn mechanical seeder had passed.

One seeder type was wheeled and manhandled down the rows, pushed from behind. Horse-drawn seeders can occasionally be found (designed for one or two horses) and with seed-bins for planting two or more rows. A marking disc was extended to one side on a pole, to mark the soil for the next pass. When horses and mules were phased off the American farm, largely by the 1940's, some of these planters were pulled by tractor until they wore out.

Cultivators—which removed weeds and piled up earth to make a stronger corn-root system—can occasionally be found, still in operating condition. In some versions, the operator walked behind, guiding with twin handles, while riding types soon became available. Different attachments meant such frames could be used for cultivating other crops.

Hand tools include the first wrought-iron hoes and the five-tined hand cultivator. Other equipment, dealing with weed-removal and with patented working edges, can be found.

The great wealth of corn-related objects begins with fall harvesting, this in two directions. One involves harvesting the entire cornstalk, the other the removal of the ears. Stalks could be cut before fully ripe, and the foliage chopped and fed to farm animals. Some stalks were stored for winter feed, and most parts could be fed to pigs.

Cut dry late in the year, stalks and leaves were chopped or shredded for bedding fodder, especially in cowpens. Other uses included mulching plants and laying bundles around building foundations for insulation.

Choppers included large, curved blades with double handles, much like a giant kitchen food chopper. Also known as feed cutters, many varieties existed to shred and pulp the stalks so they could be more easily fed. This was similar to material later stored in the towering, circular silos as "ensilage".

The plainest chopper was a narrow table with high sides, and a large, curved lever-operated blade at one end. More expensive types operated with crank and flywheel to speed efficiency. Such machines could process from 200 to 400 pounds of chopped stalks per hour. Still others had several knives for multiple cuts. The feed was not always fed raw. Farms by the 1870's began to set aside a space in the barn for large feed cookers to prepare mashes.

Corn seed itself was highly valued. For convenience in handling, the ears were generally husked or "shucked". That is, the ears were removed from the leaf-like coverings and twisted or snapped from the stalk. Two classes of tools were used for this purpose, the husking hook and the husking peg or pin.

The hook, of "U"-shaped metal, was secured to a stout leather piece which strapped to the back of the working hand. The peg was single or double pointed, with a thong attachment to several fingers. Examples have been noted with the peg of iron, antler, bone, and hardwood, usually hickory. Both hooks and pegs allowed the ears to be quickly stripped of husk.

In the field, corn stalks into about the 1940's were often bundled into individual groupings of several dozen stalks, tied with binder twine or hemp cord. These in turn were leaned together, much in the manner of wheat sheaves, but without the splayed capper top. Set up in this manner, they were called "shocks" or "stooks". To hold the larger bundles or even loose stalks together, a unique device called a corn-wrapper or shock-rope weight was needed.

A section of wooden plank, the weight had a hole, a slanting triangular end opening, and a rope. One end was knotted in the hole, and the device was whirled and thrown around the shock. this allowed one operator to pull tight the huge bundle, using the weight's opening for rope leverage. When gathered and held, the shock was tied with twine. The weight and rope were then removed for the next shock. Shock-rope weights remain almost unknown farm pieces, objects of curiosity and confusion.

Corn shellers were on every farm. Primitive types consisted of nails driven into a wood block, heads left exposed, the nails arranged to strip the kernels from the ear. Corn ears could also be shelled with a twisting motion by hand, but this was slow and time-consuming.

Machine models were more efficient, and there were two main types. One was small, of cast iron, secured by a turnscrew or clamp to a board. It had a crank-driven angled toothed disc that whirled against an ear placed in the slot. The same principle was used in the much-larger wooden frame boxed types that stood on legs with a cross-braced footer.

Aided by a large iron flywheel, these could be brought up to speed and strip off the corn in a few seconds. Shelled corn sprayed from one opening, while cobs emerged from another. Hand-cranked models eventually gave way to powered equipment when the gasoline engine became commonplace.

Shelled corn could be processed by crackers or grinders, all under the general heading of "feed mills". These could be used for making cracked corn or corn meal, or for animal feeds. Most were driven by hand-crank, and the compact metal mills could be adjusted to reduce kernels to near-powder size.

Such "hand-grist" mills were not limited to corn alone. Much farm equipment was multi-purpose, and the grinders were desired for other tasks. They could also grind rock salt, bone, shells and bark. The first was used to melt ice or make ice-cream, the next two were feed supplements, and the last shredded bark for medicine or tan-bark for leather.

Also collectible are corn baskets, many of hand-made wood splint, and shovels and scoops used to transfer ear or shelled corn. Seed corn, or corn ears for household use, could be sun or fire dried. Pronged iron racks, ceiling-hung, held the ears, but corn was also suspended by inter-twining the attached husks.

On long-ago farms, ear corn was not wasted in any way. Husks were used to stuff mattresses and make dolls. Cobs were used to kindle fires, to stopper jugs, and they made fine pipes. A favorite pioneer toy was about two-thirds of a corncob with two large chicken feathers inserted in the soft center. This made a type of helicopter or buzzer when thrown.

The communal gathering, the husking or corn-shucking bee, is part of our heritage, and colored corn was glued into folkart pictures. Even in the farm kitchen, cornbread molds and muffin pans attract interest, not to mention corn poppers. Fortunately, most corn-related items are still reasonably priced.

Values

Seeder, 18 in. tin tube, cotton seed-bag, shoulder strap . $25-35

Row-guide stakes, iron, plain, pair, 26 in. long $5-8

Row-guide stakes, for tripping horse-drawn planter, pair, springs for wire, footrest, 35 in. long $15-20

Seeder, tin bin, double wood handles, bottom 3 in. wide $22.5-30

Seeder, handled and wheeled, push-type, makes furrows $60-75

Cultivator, walk-behind, two-row, weed shovels $30-40

Hoe, hand-forged iron head, old handle, 5 ft. long $12-15

Chopper, two handles, double blades, 11 in. long $12-15

Chopper, long wood tray on legs, curved end-blade $40-50

Chopper, platform, two blades, lever-operated $50-60

Husking hook, leather palm guard, shoestring laces $7-10

Husking peg, handcarved hickory, 4 in. long $4-5

Husking peg, bone or antler, 3¾ in. long $5-7

Shock-rope weight, old white paint, 15 in. long $7-10

Shock-rope weight, heavy wood, with attached sickle-bar blade for cutting twine, 19 in. long $12-16

Sheller, "Pennsylvania", hand-cranked, 34 in. high $35-40

Sheller, metal, hand-twist type, unusual, 6 in. long $65-75

Sheller, clamp-attached, cogged side wheel, 13 in. high $20-25

Feed mill, adjustable grind, old red paint, 17 in. high . . $17.5-22.5

Feed mill, crank and flywheel, ajdustable, 24 in. high . . $35-45

Basket, covered barrel type, storage, square base, round body, 30 in. high $165-180

Basket, splinted strips, early, for stored shelled corn, hand-carved wood handles, 15 in. high $90-125

Scoop, hand-carved wood, for ear corn, 38 in. long $80-95

Doll, corn-husk, linen apron, 8 in. high $15-20

Cornbread mold, tin, 8x12 inches $7-9

Corn muffin pan, cast iron, six compartments, 5¾x9¾ inches $26-32

Corn muffin pan, cast iron, three legs, handles, 7½ in. diameter $50-60

Corn popper, wood handle, wire basket, lidded $11-14

Cottage Cheese
(See also Cheese-Related)

Associated with cottage cheese making are a number of highly desirable collector pieces. Made from skim milk, curds and whey were processed to become the familiar soft white cheese. Cottage cheese is also called "Dutch" cheese, and "pot" cheese. Sieves and collanders are the most sought-after pieces.

Values

Cottage cheese collander, redware, clear interior glaze, single handle, 4 in. high	$70-90
Cottage cheese collander, spatterware, glazed interior, handled, 13 in. diameter	$145-160
Cottage cheese sieve, tin, cylindrical, matching whey-pan, punched designs, 5 in. high	$110-135
Cottage cheese sieve, tin, heart-shaped, 4¼ in. high	$40-50
Cottage cheese sieve, tin, round, 6 in. diameter	$40-50
Cottage cheese sieve, heart-shaped, punched tin, 6x7½ inches	$82-95
Cottage cheese strainer, woven splint, old paint, 13 in. diameter	$305-350

Cream-Related
(See also Milk-Related)

Cream was the most valuable part of milk. After it was taken off, the nonfat or "skim" milk was sometimes simply fed to the pigs, or made into cottage cheese by thrifty farmfolk. Though a smaller number of cream than milk items are available, they are all of interest.

Cream could be taken from settled milk by dexterous use of a solid-bottom skimmerhorn or tin skimmer, but some of the light cream was always wasted. The better method was the cream separator, the gravity and mechanical models. The simplest was a tin cylinder on legs, with the bottom narrowing funnel-like to a brass or galvanized metal spigot. Many types held about 10 gallons.

Sometimes on the side, always on the bottom, a long, slit-like window let the operator know when the cream had been reached so the tap could be turned off. Such devices both settled the milk (let cream and water-milk come apart) and allowed them to be put in different containers.

The later mechanical separators were large and expensive, but were worthwile if the farmer milked many cows. Most were handcrank operated, and skim milk and cream separated (by specific gravity) to emerge from different spouts. Most large separators are considered more decorator objects than interior collectibles, at least by the average farm-item collector.

Cream-settling cannisters, called "setting cans", exist in many sizes, with 10 to 20 quart sizes popular. A number had long upper side windows so the cream level could be seen. Some could be set on the milk-bench, while others had handles for carrying. As with milk cans, some were brass-plate marked as to owner. Many were made of tin with a close-fitting cover and wire-bail handle.

While more directly associated with the farm kitchen, cream whippers of many kinds exist. Some are small combination whippers and egg-beaters, often fitted over a glass container. Others were designed mainly to whip cream, and some would make up to 2 quarts. Most were crank-driven to move whipping paddles at the necessary high speed.

Values

Cream settler, cannister, top side window, 16 in. high	$12-16
Cream container, tin, bail handle, cylindrical, 15 in. high	$10-14
Cream skimmer, horn, old, 5¾ in. long	$12-15
Cream separator, wooden legs, painted tin body, window, brass on/off spigot	$35-45
Cream pail, copper, iron ears, wire bail handle, wood handgrip	$50-60
Cream funnel, tin, old, 6 in. high	$9-12
Cream whipper, "A. & J.", bowl type, ca. 1900	$11-14

Crocks

Stoneware or heavy baked clay and glazed-surface containers, served many farm functions. Sizes ranged from one to ten and more gallons, with gray and cream colors quite common. Larger crocks tended to have handles, either applied near the top, or supports for wire bail and turned wood handles.

Most also had lids or covers, though some of these have now disappeared. Smaller examples were also of stoneware, while larger crocks had wooden covers. Crocks were used for brining meats, making sauerkraut, and turning cucumbers into pickles. Being heavy and somewhat unwieldy, they were yet superb storage containers and kept all manner of food.

Smaller, plainer crocks are still available in quantity. Large crocks with decorations or a maker's mark are eagerly sought, and prices steadily rise. The following price-ranging structure assumes near-perfect condition.

Values

Crock, cobalt blue feather motif, half-gal. size	$49.5-65
Crock, six-gal., incised cow in blue, 13 in. high	$55-75
Crock, three-gal., quillwork decoration, 13½ in. high .	$75-90
Crock, "New York Stoneware Co.", 8 in. high	$105-125
Crock, ten-gal. size, comic sketch of old woman with curly coiffure in cobalt slip, Albany slip interior, 17¼ in. high	$595-675
Crock, four-gal., 'Brady and Ryan", 11 in. high	$190-225
Crock, three-gal., cobalt hen pecking corn, 10½ in. high	$165-185

Dippers

Long-handled and cup-ended, dippers were used to lift up water from the springhouse pool and milk in the dairy or milkhouse. Most were casually made, for drinking or for transferring a liquid. Large numbers were made of tin. There can be some confusion with ladles, though the latter were usually associated with cooking of some kind.

Each could be made of "better" materials, like iron and "poor man's gold", brass or copper. Some dippers had predetermined volumes and served as measures. Very early dippers were no more than an open-sided gourd. In fact, the night-sky's "Big Dipper" was once known as the "Drinking Gourd".

Values

Dipper, iron and brass, handle signed and dated 1844,
    19 in. long $140-160

Dipper, tin, Cincinnati maker-mark on side, one-qt.
    capacity, hollow handle $14-18

Dipper, wood, primitively carved, 10¼ in. long $17.5-23

Dipper and strainer, pair; wrought iron, 20½ and 21½
    in. long. The lot $60-80

Dividers

A compass-like device, this instrument divided lines and
transferred measurements. It was necessary for making new
equipment parts based on originals so that certain farm repairs
could be made.

Some early dividers are handmade of wood, but metal types
are more common. Arms are iron or steel, and some examples
have brass scales or joins. Collectors look for maker-marked and
unusual dividers, and those with brass.

Values

Divider, wrought iron, early, 6 in. high $5-7

Divider, cast iron, brass joint, 10 in. high $10-15

Divider, "Lufkin Rule Co.", 9½ in. high $15-20

Divider, "Peck, Stow & Wilcox Co.", brass joint, 12 in.
    high $20-25

Divider, "Pexto 8", 8 in. high $14-16

Doorstops

Heavy, compact weights associated with doors were not
always used only in the farmhouse. Doorstops—either the
common carpet-covered bricks, or the more collectible ornate
cast-iron objects—were used elsewhere on the farm.

Any door normally has two positions, open, and, closed.
Doorstops allowed variations, as a door could be secured at any
desired opening size. As the name suggests, doorstops prevented
a door from slamming shut with force and possible damage to
people, pets or door "lights" or the glass.

Dipper, heavy tin, maker-stamped in side, 13 in. long. $10-14

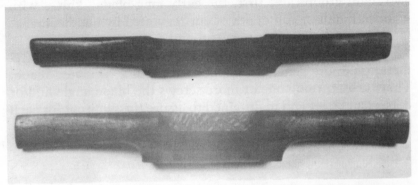

Drawshaves, top example 9 in. long. Top, iron blade, one-piece hardwood body.
$14-18

Bottom, beechwood drawshave, good surface polish, brass platform. $21-27

Regulating door-opening size was once the same as controlling the flow of air in a building, affecting temperature and draft. A farmer might wish to hasten grain drying by holding open the granary door, or use a doorstop to help "air out" the dairy house. Less affluent farmers either used makeshift weights or a wooden wedge between the door and floor.

Values

Doorstop, cast-iron bulldog 6 in. high $75-90

Doorstop, cast-iron, lady with bouquet, 6½ in. high . . . $22.5-30

Doorstop, cast-iron, eagle and 13 stars, 6¼ in. high $75-95

Doorstop, cast-iron cat, 9 in. high $60-75

Doorstop, cast-iron ram, 9 in. long $45-55

Doorstop, *pair;* cast-iron parrots with old polychrome
    paint, 6 in. high $75-90

Doorstop, cast-iron, rabbit, old white paint, 10½ in.
   high $110-125
Doorstop, cast-iron, basket of flowers, old polychrome
   paint, 5¾ in. high $25-30

Drawknives

Wood-shaving tools with a blade in the center and handles at each end, drawknives are pulled or drawn toward the user. The two main groups are, first, the early long-bladed knives with wooden handles, and depth of cut regulated by handle angles. Good work requires both control and strength.

The second class, especially the later all-metal or mostly-metal types are generally smaller and with a regulated blade. That is, by turning one or more screws the blade angle and/or depth can be set and maintained to assure the same cut for each pass.

Values

Drawknife, early, wood handles, 16 in. wide $10-13

Drawknife, adjustable handles, 15½ in. wide $15-20

Drawknife, folding handles to form blade guard $14-19

Drawknife, cooper's, blade 18 in. wide $30-35

Drawknife, "Hart Mfg.", 18½ in. wide $16-20

Drawknife, "Eskilstuna", 14½ in. wide $30-35

Drawknife, turned walnut handles, 9 in. blade $13-16

Drying Racks

"Laying something by" was always a big consideration on the farm, and it meant preserving food for the future. There was once a purpose-built drying shed for this, with sliding trays for dehydrating sliced fruits or vegetables. In warmer parts of the country, air-drying was fine, but elsewhere slow-burning fires or charcoal were used.

In pioneer days, meat was "jerked" over a fire or in the full sun. Until quite recently, orchard fruit slices were dried on racks or the roof for cold weather use. Sliced apples were a favorite, and many types of full-exposure trays were used.

Values

Herb dryer, wood, frame with narrow slats, 27x32¼
    inches $40-50

Herb drying basket, splint, wrapped risers, sliding cover,
    rare, 12¼x22x28¾ inches $225-250

Herb drying rack, wooden frame with net shelf, 29 in.
    long, 30 in. high $295-350

Vegetable or fruit-drying rack, wire-screen bottom,
    wood frame, 18x30 inches $25-30

Drying cabinet, wood, with a dozen pull-out drawer
    trays, 24x30x32 in. high $80-110

Drying tray, splint latticework bottom, rectangular,
    19x29½ inches $180-220

Drying basket, wood splint, 20½x23 inches $35-45

Egg Collectibles

Eggs were one of the first items to be used and traded on the American farm. They were so common that eggs may have coined the phrase, "a dime a dozen", which in fact was the price in the late 1800's. Eggs were eaten in many ways, used as Easter symbols, and hatched in "clutches" to produce offspring. All farms had chickens, hence, eggs, and the assorted equipment necessary.

Egg candlers were widely used. This was often a tin box with a hole and a candle inside. Better versions were of painted or japanned metal, with a window covered by a thin sheet of split mica, once called "Icing-glass". Such candlers had an oil or alcohol burner, and eggs were held against this light to determine the interior quality or whether they were fertile.

Scales were used to determine egg weight, hence class or grade and value. Some had springs, others a balancing arm with a light sliding weight. One of the more intricate had a series of flip-up leaves marked in ounce fractions, which gave the final weight.

Packing for storage or shipment was very important. Though eggs are themselves nature's best-designed package, they remain fragile. Market carriers were of heavy cardboard, while the square wood-slat box-like crates were widely used. These contained one gross eggs in four layers, though bigger types held

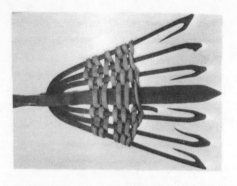 

Left, Eel-spear, hand-forged head, wedged and hooked tines, wood-splint interweave, about 1 foot high. $60-75
Courtesy private collection
Right, Egg-shipping crate, for mailing fertile eggs, six-dozen capacity. $22-28

more. Smaller metal mailing boxes with fertile eggs were sent from hatcheries to farms.

Egg-collection baskets are synonymous with farm life. They began as small- to medium-size wood-splint baskets and progressed to cheaper heavy-wire containers. Some had raised bottom centers to distribute weight better. Almost any compact basket with a sturdy handle has been termed an egg basket, especially the smaller twin-lobed splint types.

Directly related are the many kinds of artificial eggs made to encourage a hen to lay in an otherwise empty nest. Such "nest-eggs" or "hen-set" eggs are of two common sorts, wooden and handpainted white or of blown milk glass. The last are the most desired by collectors, though cheaper mass-produced pottery eggs can be found.

Values

Candler, early tin, interior socket, 8 in. high..........	$20-28
Candler, mica window, kerosene burner, black	$12-15
Scale, egg, pointer on side-face, ca. 1930	$7-10
Scale, egg, arm and weight, metal egg-cup	$6-9
Scale, egg, flip-up leaves, yellow paint, metal	$11-15
Egg crate, wood-slat, carrying handle, for four layers of 36 each, sliding top	$30-40
Egg container, metal, for shipping fertile eggs, holds 48 eggs, hinged top...........................	$22-29
Egg basket, buttocks-type, splint, 3½x6x7 inches......	$75-90

Egg basket, willow wicker in five colors, 13 in. high ... $80-90

Egg basket, splint, child-size, 8 in. to handle top $85-95

Egg basket, heavy wire, round, factory-made......... $9-13

Egg basket, thin wire, collapsible, 6 in. diameter $35-45

Nest-egg, wood, white paint heavily pecked $9-13

Nest-egg, blown milk glass, white, 2⅞ in. long $45-60

Nest-egg, quality pottery resembling porcelain $25-40

Eggbeater, cast iron, "Dover", wood handle, tin
    blades $12.5-14

Eggbeater, tin, "Pat'd 1923", tin gears, wood handle... $10-12.5

Farm Conveyances—Horse-Drawn

Throughout much of rural United States, following the Civil War and until about WW-I, a wide variety of horse-drawn conveyances were used. Though different, all were appropriate to time and place in terms of intended use.

Farm conveyances can be placed in seven large categories, some overlapping, and there are off-shoots that do not conform to the main descriptions. These are: Hay wagons, farm wagons, spring wagons, buggies, surreys, buck-boards and road carts. In an attempt to relate their old-time use to what we presently know, each brief description is summarized with a comparison with a vehicle on the road today. This is termed "Modern Counterpart", abbreviated "MC".

The hay wagon was simply a sturdy chassis with a long tongue, capable of being pulled by from two to four horses or mules. It had a rectangular hardwood bed, and generally stayed on the farm. Sides could be raised by adding boards or poles to holes along the edges, and it brought in everything from hay to sacks of grain.

A hay wagon was so all-purpose it could even be used to move small outbuildings. (Anyone who "socially" took a hay wagon to town could be accused of either poverty or poor taste.) These wagons seemed never to wear out, and when tractors replaced horses, iron-bound spoked wheels were often replaced with pneumatic rubber tires. MC: Metal farm wagons.

The farm wagon, often locally made by skilled labor, was a refinement of the all-purpose wagon, being generally more narrow, with painted sideboards. It could be used for hay and

77

field-crops, with the appropriate extensions, but it remained slow and heavy.

A major purpose was freight transportation to the nearest population center, perhaps taking in raw milk and bringing back ground feed and provisions. Some farm wagons had hand-set brakes to make uphill trips less hard on horses and make downhill ventures a bit less exciting. MC: Older U.S.-made pickup trucks.

Spring wagons refer to the steel leaf-springs under the seat, giving a smoother ride; for the first time passenger comfort was considered over freight-carrying capacity. Two-seaters were popular, as were single-seaters with the rear devoted to cargo space. Seating ranged from plain upholstered seats to plush easy-chair styles that sometimes had a high, flat weather-roof.

As with the farm wagon, simple cast-iron steps were provided. Most had painted or japanned wood, and stood high on wheels with slender spokes. These combined a certain elegance with utility. MC: Small imported pickup truck.

Buggies were the runabouts of their time, the one-horse jump-in-and-go conveyance. Still used by the Amish peoples today, they did not use an all-enclosed cab, but a few boasted a fold-up, convertible top. For the first time, and a great step forward from saddle riding, the passenger compartment was of primary importance.

It was comfortable, even plush, and there was little room for loads other than personal belongings or baggage. The buggy was designed for people to meet people, and many a male went courting in a similar buggy. MC:Imported, convertible sports-car.

The surrey could be called an enlarged, elongated buggy, and it had both front and rear seats. Many were even fancier than buggies, and all had provisions for tops. In general, two kinds of tops were available. The common leather-covered extension-type was slightly curved, enclosed in the back, with supports that could be removed to fold the top.

The canopy top, often fringed, was either flat or slightly convex to shed water. Many types were of heavy fabric, sometimes rubberized. Surreys provided a touch of class, and were favorite vehicles for the traditional Sunday-afternoon drive to visit people and places. MC: Large luxury automobile.

While buck-boards are generally associated with regions west of the Mississippi River, they were yet used elsewhere in the late 1800's and early 1900's. The buckboard was a type of light delivery wagon, quite plain and business-like. Most had long shallow bodies with sideboards, and the simple low seat was placed very near the front. They were popular because if families had only one road-vehicle, the buck-board type carried both people and a fair amount of cargo. MC: A small delivery van.

Road carts were two-wheeled no-frills personal conveyances for one-horse use. They were ideal for road-racing, and a few were fitted with a plush bucket seat; they were thus similar to the much earlier "shay". Some could be fitted with a buggy-like top, or under-seat storage for small items. MC: Motorcycle and sidecar.

Those who collect horse-drawn conveyances have, presumably, solved the problem in common with those who collect large agricultural machines. Not only must considerable time be put into restoration in some cases, but a large storage-display area must be found to protect the pieces from the weather.

Values

Hay wagon, iron-rimmed wheels, hardwood bed with "stakes"	$75-100
Farm wagon, sideboards with stencils, orig. equipment	$150-200
Spring wagon, faded stenciling, upholstered seats	$300-400
Buggy, convertible top, doctor-owned, mass-produced	$350-450
Buggy, runabout rig, open-topped, rubber-tired, "name" maker	$500-650
Surrey, "Deering Weber", two-seat, orig. leather seats, all parts, gas lights	$3500-4000
Buckboard, fine condition, with one family for over 50 years	$350-450
Road cart, most of orig. leather harness, average-good condition	$175-225

Feed Boxes and Scoops

One of the more important barn storage units is called the

feed box or grain chest. It was used to store shelled, ground or mixed feedstuffs for barn animals and served as a distribution center. The box had several bin-type compartments, each with a different feed.

Of wood, some boxes were poorly built, but still tight enough to prevent ground feed from sifting out and keep mice from entering. Each bin held several bushels so that frequent filling was not required. Construction was a rectangular form, and a typical size might be 2 feet deep, 4 feet wide and about 4 feet high.

A typical feed box had a top with partial slant-lid that raised from the front. This allowed easy opening and access to the contents; some lids projected an inch or more so one elbow could be used if hands held a scoop and a bucket. Some early and well-constructed (hardwood; dovetailed corners) feed boxes are considered as primitives; they can bring high prices.

Grain scoops associated with the feed boxes are popular collector items, and they exist from very old to recent, with newer types being used today. The scoops are either square or semicircular in end-on view. Older types are all-wood, later examples usually have metal sides and bottom, this of heavy tin or galvanized iron or steel.

Scoops were made in many sizes, from dry-quart to about half a peck, and two designs predominate. One has a handle projecting to the rear, secured to the wooden base of the scoop. The other has the handle projecting forward from base to lip, perhaps further connected to the scoop sides for added strength. All are designed for one-hand use.

Very old one-piece handcarved wooden scoops sometimes have a peculiar rear-projecting handle. High and shallow, the wide section is hollowed, leaving an elongated opening. Four fingers could be inserted in the opening, giving better control while in use.

Not surprisingly, many of these scoops show extensive wear, especially around the projecting lip and bottom. Many types can still be obtained at farm auctions at reasonable prices; values listed here have been noted at antiques shops and shows.

Values

Feed box, hand-sawn lumber, hardwood, three vertical
    compartments, 3½ feet high $175-250
Grain scoop, all-wood, hollowed handgrip, primitive . . $55-70
Grain scoop, all-wood, solid wooden handle $40-55
Grain scoop, metal sides/bottom, wood base and
    handle $15-25
Grain scoop, metal, top-handle of wood $8-12
Grain scoop, factory made, all metal, top handle
    metal $7-10

Feed Containers

Feeding farm animals was always of first interest to the
farmer and many containers were made. Hog troughs—first of
"V"-shape planks nailed into wide rectangular end-
boards—were on every farm. They were later made of non-
decay metal, and often half of an old water heater was cut for
the purpose, of ¼ in. boilerplate.

Sheep feeders were typically set off the ground, and early
examples made from scooped-out logs are in favor. Perhaps the
most in-demand collector pieces are the early feeding boxes
associated with horse-stalls.

The front of the stall had a deep manger for hay and forage.
To one side was a special built-in deep-walled box, where
valuable grain and mash were fed. These were usually made of
heavy woods so they could not be gnawed apart by the horses.
Collectors often obtain these solid containers from barns and
animal sheds being torn down.

Values

Sheep feeder, hollowed log, four angled legs, 22 in.
    high $16-22
Feeding trough, for hogs, two-sided, 3½ ft. long $7-11
Horse-feeding box, rectangular, 13x17x8 in. deep $15-19
Cow-feeder, metal box, for dairy barns $6-8
Feed box, cast-iron, 16 in. wide, for stall corner $22-28

Fencing Tools
(See also Barbed Wire)

Fencing served two good purposes, penning in animals and keeping animals from crops, truck garden or family garden. After wood rails, the better way was wire fencing, supported periodically by long posts. A number of collectibles remain from this aspect of farming. Often the first was the wooden reel-type string winder, holding hundreds of feet of cord; this was used to run a straight fencerow.

Post-hole digging was backbreaking work, so labor-saving devices quickly appeared. An early version was auger-like, with a short screw or thread end, long shaft and wooden "T"-shape handle. Even before that was the primitive two-man plummet, using a pointed iron tip and log body with upper crosspiece. Four other types, used from the early 1900's on, sank straight and narrow holes.

One had a single long handle, and angled shovel edge, and was used to both loosen and scoop out earth. Favorites, however, had "clamshell" working edges, which remained open for digging and closed to lift out the earth.

Two of these had parallel digging shovels, with one type having a single handle which was cut full-length down the center. The handle was pulled apart to close the shovels and extract earth. A related type also had twin shovels, but two separate handles which were round.

Still another version, quite practical, had a single pivoting shovel below a strong handle. A side lever about halfway down tilted the shovel head at right angles to remove the earth. With this digger a narrow firm-walled hole could be sunk 2½ feet.

Once the spaced holes were put in, the posts were set firmly with two other fencing tools. One was the common long-handled shovel, suitable for almost any digging or scooping task. Shovels rough-filled, but the tamper made certain the earth went in tightly enough so the post would remain straight and sturdy for years to come.

Some special-made iron tampers with oval or rectangular heads were used, but most farmers distained such "fancy" objects. Instead, hardwood tampers about 6 ft. long were common, with large and small blunt ends. These seated the earth in the narrow space between post and walls. Interestingly, in early times charcoal was used for fill-in, to retard post decay.

Following post-setting, wire fence was laid out and strung up. Providing a proper measure had been taken—usually with an encased wind-up fabric tape, or the much older set of "surveyor" chains measured in rods—the wire would be but just a bit longer than required.

Good fencing now required two key steps, tightening and fastening. Short segments were pulled up to be secured before moving to the next section. Tightening, now done by tractor power, required muscle and special gear in earlier times.

A basic type used ropes and pulleys to tighten the fence, with an energy-ratio of perhaps one foot of "pull" to one inch of "tighten". Numbers yet exist, earlier examples with wooden turns instead of iron. They have been called block-and-tackle.

The other main type is the catch-and-lever, with a long iron arm that pulls to the user and a cogwheel to catch and hold the leverage gained. Whatever the device, slippage of the steel fence was sometimes a problem, and many fail-safe holders were used, all of metal. The simplest worked on a pull-tight principle, whereby the greater the force the tighter the hold.

A special fencing tool resembled a pliers and was used to pull out misplaced staples by jaw action or a curved, pointed tip was inserted. Additionally, the fencing pliers had several scissors-like cutting edges for clipping wire. Of all fencing tools, collector's generally seek the old and the unusual.

Values

Post-hole maker, plummet, iron tip, two-man	$50-70
Post-hole maker, screw/thread tip, cross-handles at top	$65-90
Post-hole digger, slanted shovel, single handle	$15-25
Post-hole digger, clamshell shovels, split handle	$20-30
Post-hole digger, clamshell shovels, double handles	$20-30
Post-hole digger, lever-tilted shovel, one handle	$25-35
Shovel, long-handled, early, factory-made	$15-20
String-winder, reel-type, wood, homemade, 10 in. wide	$12-15

Tamper, hardwood, 5½ ft. long, blunt ends $8-12

Tape, windup spool, enclosed, side folding arm $10-15

Chain, surveyor's, wrought iron links, rare $60-80

Fence-stretcher, ropes and wood pulleys $10-15

Fence-stretcher, iron, lever and cog-wheels $12-18

Fencing pliers with jaws, pointed tip and cutters $4-6

Fence-clasper, rope-mounted, single-strand, iron, 4 in.
  long $2-3

Fencing tool, "Stetler/1-2-1900", 17 in. long $35-40

Fencing tool, for splicing barbed wire strand, 15 in.
  long $25-30

Wire stretcher and splicer, "Hayes", 18 in. long $40-50

Files

Files—ridged devices for removing excess wood for shaping or metal for sharpening—tend not to be collectible in themselves, except for very old or ornate examples. They are made today, are common on every farm, and have the ability to hide when they are most needed. Wood files are wide and heavy, while metal files are smaller, of high-quality steel and with closer-spaced cutting ridges.

The importance of files as a collectible comes from earlier times, when the scarce metal was a key resource when another tool was needed. When a file wore out, there was no easy way to resharpen it. So, instead of being thrown away, the file was made into other objects.

Some were gently altered into log-moving canthooks by farmer or blacksmith, while others, due to the similar form, became long knives of the belt or sheath type. Still earlier are the strike-a-lights, the "knuckle-duster" piece for flint and steel kits. Later, some files were made into wagon-pins, even the steel cutting edge of broad axes.

Almost always, file-made American ingenuity can be recognized by the fairly small size, the fact that the objects are high-quality steel and handwrought, plus the presence of the original file cutting face. Workstyle is varying degrees of artistry, with most quite well made.

Farm wagon with all-metal wheels, high box body; note brake in front of rear wheel.
Photo courtesy Museum of Appalachia, Norris, Tennessee

Right, Animal Feeder, made from trunk of tree, heavy peg legs.
Photo courtesy Museum of Appalachia, Norris, Tennessee.

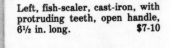

Left, fish-scaler, cast-iron, with protruding teeth, open handle, 6½ in. long. $7-10

Right, File-made objects, each made from an old steel file. Tooth-ridges are still visible on both pieces, bottom specimen 3½ in. wide.
Top, file-made pliers $12-15
Bottom, strike-a-light $40-45

Left, Very rare trap-gun; loaded with a 12-gauge shotgun shell, it was fired when the bait triggered the firing pin.
Photo courtesy Museum of Appalachia, Norris, Tenn.

Fish-spear head, hand-wrought iron, about 12 in. wide, from Maine. $50-65
Courtesy private collection

Above, Fishing bait container with sling, for minnows or dry bait, lift-off top, jappaned green, 10 in. high. $11-14

Left, factory-made fishespear head, haft broken along weld-line. Now 8¼ in. long, graceful shape. $11-15

Middle, fishspear, hand-wrought iron, pointed hafting tang, overall length 10 - 3/8 inches. Tine-barbs were put in with a thin file. $25-33

Right, fishspear head, very old, hand-forged iron head, square multi-barbed tines with outer pair held in position by an iron wedge in haft. Overall, 8 - 3/8 in. long. $30-40

Values

Canthook, hooked end, loop for long wood handle, 14 in.
    long $10-13

Knife, double-edged, bolted-on wood handle, blade
    7½ in. long $15-20

Log-dogs, with angled, pointed ends, for holding log
    firmly for sawing, 8 in. long, pair $30-40

Strike-a-light, handheld, curved finials, 3¾ in. long . . . $35-40

Fire-Starting

On the farm, fires were needed in places other than the hearth and stove. The summer-kitchen had either a fireplace or a wood-burning stove, and a slow fire was placed in the smoke-house after butchering. Some barns and other outbuildings—and areas like the workshop or forge room—also required fire. Fires in orchard "smudge pots" helped ward off early frosts.

The most common way to start a new fire was to use the old. Each night before retiring someone would "bank" the fire to preserve a bed of coals. This involved cutting off the flow of fresh air and oxygen by using a damper or by burying coals beneath a layer of wood, covered with ashes.

If by mischance the fire went cold, it could be borrowed from a neighbor if one were near. There are some special-use ember carriers, but more likely ordinary objects were more widely utilized. Then a child could be sent to retrieve live wood coals with the tin-cased footwarmer or the long-handled brass bed-warmer.

Beyond the basic and makeshift techniques, certain other objects were used to make fire, and some are highly collectible. The best-known is the flint and steel kit, the steel part also called a strike-a-light. This was piece of superhard flint and a curved metal strip designed to be held in one hand.

Often the steel had been blacksmith-made, and some examples have brass figural handles in the shape of animals. Sparks could also be struck using the back of a hunting knife's steel blade.

Flint and steel were struck together, the same principle used in some cigarette lighters today, which produced a random shower of sparks. Starting a fire this way required real skills, for the sparks had to fall into another material, called punk or tinder. Some frontiersmen swore by the pounded ultra-dry inner bark of certain trees.

In near-universal use was heavily charred cloth, which caught and held the spark. With careful fanning or blowing a portion could be induced to fire fine shavings, and a fire eventually resulted.

The three fire basics were typically kept in a container called a tinder box, which was often about five inches long, oval or square, and lidded. Large versions were kept on the farm, while smaller, flatter types were carried by travelers. These boxes were usually made of tin, well-soldered or rolled to be mostly water-tight.

The wealthy farm in early times had an expensive device which greatly resembled a small flintlock pistol. These lighters operated mechanically and produced sparks with little effort. They were more or less the lock portion of a firearm and operated by trigger. These "class" devices probably developed in the late 1700's from the wilderness hunter's occasional practice of starting emergency fires using a flintlock rifle or pistol.

The so-called firepot was used to start fireplace or stove kindling. These are small cast-iron pots, about five inches high, rounded in the middle. There is a heavy wire bail, and the lid has a semicircular opening along the rim.

A twisted-wire handle went through this opening. On the handle base was a wire-enclosed porous ball about 1¼ inch in diameter. The pot contained a flammable liquid, probably coal oil. The soaked ball was ignited and placed beneath wood or coal and provided high-intensity flame for lighting almost any material. Some, with thin brass containers, are being made today.

More sophisticated and unusual was the burning glass, which used the principle of solar energy two hundred and more years ago. Often imported, and sometimes a library reading glass or the lens from a telescope, the glass concentrated the sun's rays on a tiny area. Under proper conditions it could readily coax smoke and flame from almost any combustible material.

Early matches were also used, far different than those of today. The chemical tips could not be struck, but were used to produce flame from smoldering punk. Later came the large wooden "kitchen" matches, the strike-anywhere type, combining sulphur and phosphorous in the head.

More recent are the "safety" matches of wood or paper with part of the firing chemicals in the striking strip. Early matchboxes were tin and the matches were produced in strips or blocks.

A number of patented fire-making devices were marketed, many employing the flint-and-steel system. Some are made to this day, for lighting charcoal and camp lanterns. Some devices were simple and workable, while the performance of others would have been questionable.

One of the strangest and least-known early firing devices, referred to in the literature of the time as "phosphoric candles", was brought to America by a French scientist in the 1790's. Sometimes called the "ethereal match", it consisted of paper that had been coated with white phosphorous, and sealed by a special process within a glass tube. Shattering the tube exposed the volatile chemical to oxygen and it spontaneously burst into flame. More curious than practical, it was never widely used.

Values

Ember carrier, wood handle, flip-up lid, iron, 17½ in. long	$42.5-50
Coals carrier, heavy tin, has considerable rust, 20 in. long	$10-16
Ember carrier, wood handle, iron, 14½ in. long	$165-190
Strike-a-light, steel, finial ends, 4¼ in. long	$30-40
Strike-a-light, curved surface, brass handle of fox or dog	$95-120
Tinder box, tin, lidded, oblong, 4⅞ in. long	$25-30
Tinder box, strike-a-light, flint and tinder, very old	$100-125
Tinder lighter, wrought iron, flintlock, missing handle, 5 in. long	$255-275
Tinder lighter, flintlock, pistol grip, with brass trim, floral engraving, 5½ in. long	$750-800
Firepot, 5 in. high, lidded, with ball lighter	$45-55
Burning glass, old, 2¼ in. diameter, brass rim, loop for cord suspension	$40-50
Matches, wood, in connected strips, old wooden box	$15-20

Matches, wood strips, orig. tin box, 2x2x4½ in. long . . . $20-25
Firelighter, "Lutz Patent Fire Lighter", orig. directions
    for use, 8 in. long $45-55

Firearms
(See also: Shooting Accoutrements; Bullet Moulds; Powderhorns)

In the past two centuries, many kinds of firearms have come and gone on American farms. Use ranged from wild game hunting to slaughtering domestic animals to home protection. The two most widespread categories are rifles, shooting a single projectile from a twist-grooved or rifled barrel, and shotguns, which fired a quantity of small, round shot.

Two hundred years ago, few pioneers could afford a fine custom-made flintlock rifle made by a backwoods or village gunsmith, so nearly worn out military longarms were quite common. Later came the favorite of the frontiersman, the Kentucky rifle. Many were made in Pennsylvania, but were used in well-publicized Indian encounters in Kentucky, then called the Dark and Bloody Ground.

A stepchild of that great creation is a rural version, the Midwestern half-stock small bore caplock which can still be found in some numbers. The calibre may be .30 to .40, the barrel octagonal. Fine woods, often maple or walnut, were used for the one-piece wood portion, and the curved buttplate is usually brass.

A number have ornate patchboxes, for decoration as much as utility. Individually made, many are decorated with brass or German silver or ivory inserts of stars, animals or symbols. Such rifles were widely made in the 1840-65 period.

These, and Civil War surplus arms, served until nearly 1900 when the inexpensive .22 calibre rifle beat out most competition for a farm gun. The rifles sold for several dollars and cartridges cost only five shots for two cents. The single-shot rifle of whatever calibre or make was improved by the repeating rifle. These operated the breech action by lever, slide or bolt. Even the first examples of semi-automatic rifles are of collector interest.

Shotguns followed related development patterns, beginning with heavy muzzle-loading "fowling pieces" in flintlock. Single barrels were followed by the "side-by-side", these in turn replaced by, or converted to, caplocks. These used the metallic percussion cap.

By the late 1800's breech-loading shotguns developed, using the self-contained brass shell loading which contained primer, powder, wadding and shot. Eventually rural residents could choose either to keep the heirloom guns or obtain one of the then-new lever or slide-action "scatterguns". Just after 1900 they could even try the newfangled self-loader by Remington.

In addition to such standard firearms, often a handgun was kept on the premises for whatever need. As much as today, folks then were concerned about burglars and trespassers. Muzzle-loading pistols were kept, and by the 20th Century there was a vast selection of pistols of the revolving cylinder (revolver) type. Some were quality, some "Saturday Night Specials", but most cost only a few dollars. The new semi-automatic military-type handle magazine pistols fast gained in favor.

Unusual firearms existed for country use. One was the massive "market" or "punt" gun, used wherever large flocks of water birds were found. Up to 8 feet long and some close to a hundred pounds in weight, they were charged with huge amounts of blackpowder and shot. It was possible to fire them only from braced stands or punt-like boats, and the targets were large "rafts" of settled ducks or geese. Reports say the kickback was sufficient to thrust boat and occupant many yards rearward.

Some farms had the controversial gun-trap. These were used to frighten off, or shoot, chicken thieves and sundry other intruders. Most had a pistol-length barrel, and were triggered by a trip-wire. Set examples were secured, at say, a window, and fired in only one direction. Swivel gun-traps were designed to move with the intruder by a system of cords. All this gave rise to some interesting legal considerations, as to whether deadly force could be applied in the absence of the owner.

Values

Kentucky rifle, full stock, orig. flintlock, brass patch-box	$1000-plus
Kentucky rifle, full stock, 42 in. barrel, signed	$675-775
Kentucky half-stock, caplock, patchbox, signed maker and owner (two places)	$450-525

Kentucky half-stock, caplock, replaced hammer, German silver insets, fancy patchbox $400-500

Springfield rifle, .45-70 cal., Model 1884 .... $250-300

Winchester rifle, .38-56 cal., Model 1886, lever-action octagonal barrel $200-250

BSA rifle, .22 cal., Model 12, ca. 1910-30 .... $190-220

Fowling piece, flintlock, smoothbore, ca. late 1700's ... $700-850

Winchester, percussion, double-barrel, Class B engraving $300-350

Shotgun, "New Baker/Batavia, N.Y.", double barrel, 12-gauge $80-100

H. & R., "Topper M-48", 12-gauge, 30⅛ in. barrel $60-76

Stevens, Model 94-C, 16-gauge, 28 in. barrel $75-90

Shotgun, Whitmore patent, double barrel, 12-gauge ... $160-190

Waters flintlock conversion pistol, Model 1838, dated 1844 .... $250-270

Pepperbox pistol, Sharps, four-section square barrel ... $325-375

Pepperbox pistol, Allen & Thurber, six-section barrel .. $250-300

Revolver, Colt, Officer's Model Target, .22, six-shot, ca. 1930-49 .... $225-250

Revolver, Stevens, "Lord Gallery", .22, single shot, ca. 1898-1908 .... $225-260

Pistol, Colt, "Woodsman", .22, 6½ in. barrel $140-160

Revolver, Colt, Army .44 cal., Model 1860 with checkered grips $350-400

Pistol, Colt, semi-automatic, .32 cal. rimless $120-140

H. & R. single-shot target pistol, checkered grips $175-215

Mauser pistol, Model 1910, .32 semi-automatic, with holster $275-325

Smith & Wesson revolver, .32 cal. Army No. 2, custom grips $175-200

Punt gun—No recent examples known to have been sold

Gun trap, flintlock, handle with iron clamp, large bore, 9 in. long, rare $1200-1500

Fishing

Fishing was a major warm-weather sport on many farms, whether in creeks or rivers, ponds or lakes. Very simple gear was often used, this at a time when cane poles were 50 cents each and hooks were two for a penny. More than an agreeable pastime, a fat string of bluegills or catfish was a welcome addition to standard farm fare.

Early poles and fishing reels are quite collectible, so also lures, tackle boxes and bait-carriers. Collectors look for certain manufacturer's names, as well as items with copper and brass parts.

Values

Cricket box, tin, green paint, perforated lid and extension for fastening to belt, 3½ in. long $15-20

Cricket cage, copper, square, sheet top and bottom, screen sides, round opening with lid, 10 in. square . $47.5-55

Fishing reel, circular, fly-casting type, black metal, 4 in. diameter $11-14

Fishing reel, bait-casting type, solid brass, no gears $28-35

Fishing reel winder, all wood, drum with handle, 6x12 inches $25-30

Fishing rod basket, wicker, Victorian, oval, hinged, 40 in. long $60-70

Fishing rod holder, cast iron, "Pat. 1888", with clamp . $13-17

Fish-hook maker, very old, iron block for shaping wire to form shaft, curve, snell and eye, 8 in. high $125-150

Fish-hooks, most of orig. 50 in round wooden box $11-14

Fishing Decoys

For the discerning rural-interest collector, another type of decoy than ducks and geese comes to mind. This is the fish or fishing decoy, and they are relatively scarce. These are not lures in the sense that hooks are attached, but were made to resemble a tempting smaller fish for larger predatory species.

Hence, most fishing decoys are under about 10 inches in length, and the average might be 6 or 7 inches. The body is made of wood, sometimes select hardwoods, but almost any grade appears. In general, the careful handiwork given to the above-water duck decoys and shorebirds does not appear in fishing decoys.

Either smoothly or crudely handcarved, some bodies are

juttingly angular, others rounded and graceful. Some look like the foodfish they were made to mimic, but a few examples are more a folk-carving of what the maker thought a fish should be.

Fins are always included, are made of tin or wood, and are located in about the right anatomical positions. No mouth is usually indicated, and gills are not outlined; eyes, if present, might be of paint or bright tacks.

Fishing decoys are mostly a light color, the body one shade, the fins and tail sometimes another. Some two-tone bodies exist, with different colors in the head or tail region or in the under-belly area.

All authentic decoys will have an attachment point, an eyelet or screwed insert atop the body, forward near the head and near the pectoral or horizontal fins. A line secured to this eyelet permitted lifelike movements underwater. It should be noted that the decoys, being largely of wood, were weighted to sink easily.

These interesting objects were used in colder regions of the country in the sport of ice-fishing, both by farm and urban sportspeople. The decoy was dangled in the water beneath a chopped hole, sometimes from an ice-fishing shanty that gave the operator some protection from the cold.

The decoy was moved by a line and the motions were designed to attract pike, pickerel and walleyes. When the large fish moved in to investigate or attack, it was impaled by a hand-held fishspear.

In shallow waters old-time sportsmen would often drop broken porcelain plates into the water and the pieces settled to the bottom. This formed a brighter background and large fish could more easily be seen against this in the low-light winter conditions.

Fishing decoys are now commanding high prices, and of course, fake or altered or recent specimens are sometimes passed off as originals. Style or size are not really indicators, as many were one-of-a-kind creations, perhaps carved by the person that used them.

Beware of the decoy that has a line attachment point at the front tip area, as this is likely an old wooden lure with the single, double or treble hooks removed. Tin tail or fins should be somewhat spotted or rusted from repeated contact with air and water.

Look also for new-appearing line eyelets and fins, or fresh paint. On good decoys, the paint should not crack or alligator, because actual use took off the paint and left a thin residue of the original coat or coats. This will be smooth if the body was sanded.

Fishing decoys are fascinating reminders that humans wishfully believed they could outsmart fish. And judging from some of the well-made, lifelike decoys, they sometimes succeeded.

Values

Decoy, wood body, metal fins, 9½ in. long $30-40

Decoy, tin fins, worn paint, 13¾ in. long $60-80

Decoy, primitively carved and painted, tin fins, 9 in.
    long $45-55

Decoy, carved wood, catfish with whiskers, unusual,
    9 in. long $210-250

Flashlights

By 1900, technology was available to provide low-cost dry cell "hand lamps" (flashlights) for farmspeople. They were rather prototypical, but they did work. While the hand-held type could be had, well-dressed "swells" carried the compact vest- or coat-pocket models.

Beginning flashlights had a casing of rolled cardboard, though stronger substances were eventually used. Various metals and sheet copper came into being, and the early plastics soon arrived. As electrical flashlights became better and cheaper, they began to be used for late-night farm chores and such diversions as hunting.

Values

Flashlight, cardboard tube, convex lens, 8 in. long $13-17

Flashlight, 4 in. high, coat-pocket weight, small bulb . . $12-16

Flashlight, copper body, focused beam, 9¼ in. long . . . $18-22

Flashlight, black plastic, Winchester-marked, two-
    cell $17-25

Flax

Pioneer farmers raised an important crop in the form of the towering flax plant. It is of the genus *Linum,* from which comes our word for its major by-product, linen. Objects made from it had a pleasing "flaxen" color, were durable, but somewhat rough.

The flax was cut and stripped of leaves, for only the long stalk contained the desired fibers. These were first crushed and softened, using two devices called flax-breakers. One was the hand-held breaker of carved wood, a true primitive, that resembled a straight sword with blunt edge.

Another breaker was the mounted machine which used a long arm to crush and bend the stalk as it was fed through. It too was of wood, had one or more breaking edges or blades, and stood on four slanted legs.

In flax preparation, the next step was to take the broken flax to the hatchel, or flax-comb. This was a close-set, usually rectangular, grouping of upright iron teeth that were firmly mounted in a flat wood base. The flax was pulled through until the strands were straight and all debris removed.

Loose flax was then secured atop the flax wheel, to be spun into thread. This wheel greatly resembles the wool-spinning wheel, and many of them are extant today. While thought to have been used for wool, they actually made loose flax into thread.

Once thread was obtained, it was wound into spools or balls for making into linen by various weaving processes. All-linen creations—except in the wealthiest farm homes—were rare and only a bleached Sunday-dinner tablecloth might be produced. Instead, the universal early textile fabric was a mix of linen and wool to combine strength and softness. This was the famous linsey-woolsey.

Shirts and "britches" (breeches or trousers) were made from linsey-woolsey, and it was said to "itch like the Devil but wear like iron". It's popularity was based on there being nothing else available.

Values

Hatchel, forged teeth, wood base 9x17 in. $30-40

Hatchel, base decorated with hearts and birds, dated
 in mid-1800's, wood box-cover for teeth $135-150

Flax-breaker, 19 in. long, hand-carved wood, handled . $40-50

Flax-breaker, floor-mounted, all wood, pegged, single
 breaker arm, 43 in. high $100-120

Flax wheel, still with flax remnants on upright holder,
 all parts present, fine condition $225-275

Linsey-woolsey apron, some wear and damage........ $9-13.5

Linsey-woolsey boys' shirt, very early, butternut stain
 or dye....................................... $30-40

Linen table-cloth, early, plain, 26x42 in.............. $45-55

Linseed cake animal-feed container, factory-made box . $10-15

Linseed oil can, early, lithographed label, 7 in. high ... $8-12

Linen press, miniature, in wormy chestnut, two drawers,
 for handkerchiefs, 15½ in. high $515-600

Flax winder, eight-armed, wooden pegged, 4 ft. high .. $150-175

Flax basket, woven splint, wood handles, 6½ in.
 diameter $40-50

Fly Control

Some early drovers—profesionals who guided large herds of
animals to market cities over land routes—were said to be able
to flick flies from animals with the long "blacksnake" whip.
Most rural people did not possess such an esoteric technique,
and countered flies with other devices.

Flies were a real problem. They annoyed, were unsightly,
and spread disease. To control them, screens and nets were used
in the warm months. Fly whisks were employed, and an early
example borrowed on nature, being a strip of horsetail nailed to
a short pole.

Swatters, wood or wire handles with fabric, rubber or wire
flaps, were used in numbers, as was the sticky ribbon or "fly
paper". These uncoiled from a cardboard tube and were hung
in special places, like the milking parlor.

Fly-traps were of different designs, ranging from glass to
wood to wire screening. All were on a similar principle, attract-
ing flies with bait and then trapping them in an enclosed area.
Sprayers were also widely used when non-toxic chemicals
became available.

Values

Flywhisk, made of horsetail and stick, primitive....... $5-7

Flywhisk, twine and turned wood, handle 18 in. long .. $6-9

Swatter, flat wood handle, advertising, "Swartz'
    Cure".. $3-5

Swatter, homemade, innertube flap and wood handle.. $1-2

Trap, metal screen, cone-shape, 6 in. high $25-35

Trap, glass, fits under fruit jar, 5 in. diameter $15-20

Trap, wire and wood, primitive, 2x3x12 inches $25-30

Sprayer, tin, half-pint reservoir, rusted $2-3

Sprayer, all brass, wood plunger handle, quart reser-
    voir $30-35

Fly-catcher, tin and wire, 9½ in. long.............. $15-20

Folk Art

One of the more collectible areas relating to rural activities is the category known loosely as "folk art". The area is somewhat related to primitives and Americana, although few persons can agree on exactly what the term means.

Generally, the folk art items are handmade or hand-carved of wood, and were done by nonprofessional workers or artisans. The items may be in a known field, such as early decoys or shorebirds, or may be one-of-a-kind.

Folk art pieces are not generally useful, however, but are decorative renditions of whatever the maker desired. Animals and people are often depicted, and such works generally evidence a charming simplicity and a stiffness as if frozen in motion.

Folk art is what the name suggests—artwork turned out mainly by common folks who developed whatever talent and skill on their own. Sometimes the objects are further decorated with fabric or bases, but painting was also common. Large and dramatic folk art pieces are much in demand at present.

Values

Figure, of bearded man, primitive details, 20th Century,
    12½ in. high................................. $30-40

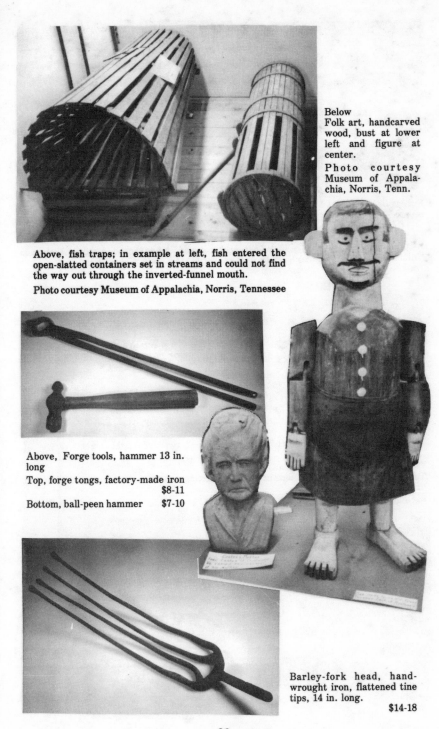

Below
Folk art, handcarved wood, bust at lower left and figure at center.

Photo courtesy Museum of Appalachia, Norris, Tenn.

Above, fish traps; in example at left, fish entered the open-slatted containers set in streams and could not find the way out through the inverted-funnel mouth.

Photo courtesy Museum of Appalachia, Norris, Tennessee

Above, Forge tools, hammer 13 in. long

Top, forge tongs, factory-made iron $8-11

Bottom, ball-peen hammer $7-10

Barley-fork head, hand-wrought iron, flattened tine tips, 14 in. long.

$14-18

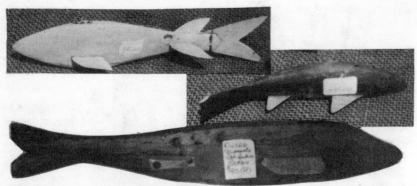

Fishing decoys, about 5 and 6 in. long, respectively, both with in fins. Top, light-painted tin tail. $45-55

Middle, fish with unusual curved and carved tail. $40-50

Bottom, fishing decoy, hand-whittled with tack eyes and tin fins tacked to body. Label reads "Minnesota ice-fishing decoy". With wooden, curved tail, item is about 8 in. long.

$50-65

Left, footwarmer, pottery, handled, for hot water, "A Warm Friend".

$35-45

Forks; left, hay or straw fork, four tines.

$8-11

Right, short-handled barley fork, two-tines, lightweight.

$12-16

Forge tools, blacksmithing hammers.
Top, square-face iron head, unusual iron reinforcing rod handle. $5-8

Bottom, hand-forged head, hardwood handle, 9 in. long.

$8-11

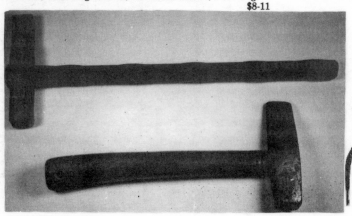

Figure, made from tree fork, jointed arms and legs, tree-
    bark trousers, button eyes, fur mustache and eye-
    brows; hair missing, 60 in. high $710-850

Oxen, pair, wooden, painted, 11½ in. long $50-65

Rooster, black and red paint, 10½ in. high $315-400

Dog, carved, pine, excellent detail, primitive, cross-
    hatch coat, brown patina, 15½ in. long $750-900

Turkey, carved wood, red and white paint, original
    varnish, 8 in. high $155-175

Seascape, carved and applied land masses with boats,
    houses and lighthouse; old polychrome paint, 6x40¼
    inches $325-400

Lid, from grain bin, pine with orig. red paint and
    stenciled black horses and hearts, 20½x37 inches .. $195-225

Black boy bust, missing teeth, 6¼ in. high $85-105

Box, decorated, inside cover with five-pointed star, pine,
    5¾x7x10¼ inches $80-90

Carving, ca. 1890, country character with whiskers,
    2¼ in. high................................. $15-20

Carrousel horse, leaping, leather tail and ears, 6 in.
    high $180-200

Monkey, eating coconut, carved, 9¾ in. high......... $90-115

Elephant, primitive style, 3x5½ in. high $32.5-40

Swizzle stick, four graduated hoops, spun between
    hands...................................... $45-55

Footwarmers

There might seem to be no place for footwarmers on a farm,
but they were not only practical, they were necessary. In horse-
drawn days the warmers were needed for cold-weather buggy
or surrey driving, the former a smaller version of the latter.

Too, much winter travel took place, and the warmers helped
make long trips possible. Placed under the feet, rising heat was
trapped by the heavy lap-robes and helped keep the entire body
warm. Early robes were hides until the development of the
heavy woven robes.

Some footwarmers in the early 1900's were referred to as
"carriage heaters" and were basically a metal frame with ember
drawer, with a rug-like cover to prevent burns. In cross-section,

most were oval, triangular or rectangular. Before the factory product, warmers were a box-like iron container with coals shelf, the whole contained in a light wooden frame. Many varieties were made, and they are considered primitives today.

Later types included pottery containers that had a plugged hole and held hot water. Related but fancier were the warmers with a thin copper body and copper or brass screw-top secured against loss by a chain. Boiling water was poured into both types, and they held heat for hours.

The widest-used farm footwarmers were undoubtedly the omni-present soapstone or steatite heaters. Square to rectangular, with a heavy wire bail at one end, these were heated on or near the stove before venturing into the cold. The material does not crack under high temperatures and retains heat for an amazing period of time.

Values

Carriage-heater, heavy metal, rug-fabric cover, coals
drawer, handle, 18 in. long $30-35

Carriage-heater, two-person size, oval, rug-covered with
floral designs, 20 in. long $35-40

Ember-box, old, wrought coal-chamber and tray, perforated metal, turned wood frame, bail handle . . . $90-125

Pottery footwarmer, cream-colored, keg-shape, no
maker-mark, 16 in. long $40-50

Copper footwarmer, brass screw-top, cylindrical, 15 in.
long $37.5-45

Footwarmer, soapstone, one inch thick, 8x12 inches . . . $12-15

Footwarmer, soapstone, 1¼ in. thick, 8½x13 inches. . . $14-17

Footwarmer, square tin perforated box in wood frame,
bail handle, 9½ in. high $125-150

Footwarmer, pierced tin box, wood frame, turned posts,
9x11 inches $80-100

Footwarmer, tin, fancy, interior coal-pan, 5¾x7½x10½
inches $110-125

Forge Tools

Though the professional blacksmith (iron-shaper) and whitesmith (iron-finisher) had a more elaborate array of tools, the farm forge room was hardly short of tools. Basic implements were there, to support the forge, itself rather small.

Farm forges began and remained quite simple. They were first stone-walled and lined with fire-brick. By the late 1800's they were cast-steel and hooded, to carry off acrid fumes. Air was pumped into the base by fans or blowers powered by cranks or levers with flywheels. The earliest examples had attached giant bellows of oxhide.

Almost inseparable from the forge was the shaping surface, the anvil. From 25 to 150 and more pounds in weight, anvils were securely mounted, often on a solid section of tree trunk. Each anvil had a flat rectangular top, and a pointed end for rounding. Most had squarish top holes for adding special inserts for cutting and shaping tasks. Such inserts are called "hardies".

Forge tools are of two main types. There are those used to pull glowing iron from the forge, and those used to pound the pieces into shape. Smaller tongs held and turned the iron while it was being worked by hammer.

Forge hammers, in assorted sizes, had a round or rectangular surface on the head, opposed by a straight edge or smaller flat surface. Such hammers could be used to flatten and cut the iron, while specialty hammers like the lightweight riveting hammer were used for that purpose. A typcial forge hammer had a head weight of about 3 pounds.

Various long-handled iron tongs placed iron in the forge, repositioned it, and removed it when the right temperature was reached. A good forge worker could judge the degree of heat simply by the color. The iron was then worked on the anvil, the piece usually held with short-handled tongs. And, if a farmer lacked a certain forge tool but had sufficient skill, he could simply make one.

Values

Forge, hooded, lever-action airpump, 50 in. high $150-190

Forge, open, handle-crank airpump, 39 in. high $125-160

Anvil, 125-lb. size, with stump base $90-130

Anvil, 50-lb. size, without base, hardy holes $75-100

Anvil, small, 14 in. length $35-45

Tongs, 24 in. handles, nipper tips $9-12

Tongs, 19 in. handles, flat-jaw type $7-10

Tongs, 21 in. handles, grooved tips $8-11

Hammer, 3-lb. head, wood handle.................. $10-14

Hammer, 1½ lb. head, wood handle $8-11

Hammer, ball-peen head, wood handle.............. $7-10

Hardy, cutter top, for anvil insertion $4-7

Bolt cutter, handled, double-action clip $30-40

Hardy, flat top, fits anvil hole, for shaping $3-6

Hardy, rounded "fuller" top, 2½ pounds weight $4-7

Forks

Forks—tined metal heads with long, fairly straight wooden handles—are the classic farm implement. They exist with from two to ten tines, and aided in haying and harvesting, handling animal wastes, digging and scooping.

Hay forks are the "pitch-forks" used for moving piles of hay from one place to another, plus loading and unloading. The most common was "the" hay fork with three tines, but one with four was also widely used. Hay forks, to differentiate from similar types, have the outside tines with gently sloping shoulders to aid in pitching or releasing the hay. Four-tined forks were also quite useful in handling the shorter "slippery" lengths of grain-straw.

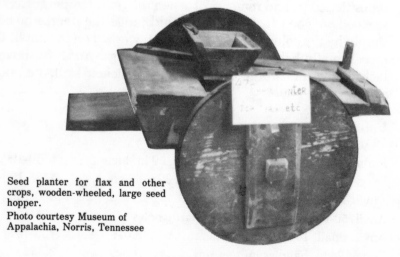

Seed planter for flax and other crops, wooden-wheeled, large seed hopper.
Photo courtesy Museum of Appalachia, Norris, Tennessee

Manure forks, designed for a mixture of animal waste and straw or shredded corn fodder, had from four to six tines. These were wide-shouldered on the outside base. The heads were designed as much for digging and lifting than for pitching, except for the one type with tines at right angles to the handles. This was used to rake manure (muck-raking) and pile it. Manure gathering cleaned out the animal pens for fresh bedding, and the manure went back to the land as crop fertilizer.

Barley forks had long tines to securely grasp the sheaves, and alfalfa-hay fork-heads were long and wide, generally four-tined, to pick up the short alfalfa stalks. Still to be found is the two-tined lightweight grain fork designed to be used all day during threshing.

The short-handled multi-tined field-vegetable fork had flat surfaces on the tines and were used to load and screen beets, onions and potatoes. A related type was used to transfer coal and coke, but it had straighter tines and a squared instead of rounded head configuration. The spading fork, still available today, was used to turn soil and had the four tines flattened on the front or turning surface.

Values

Hay fork, three-tine, 6 ft. handle	$9-12
Hay/straw fork, four-tine, 5½ ft. handle	$8-11
Manure fork, five tines, hickory handle	$6-9
Manure fork, four tines at right angles to handle	$8-11
Barley fork, four tines, wide-spaced	$12-16
Alfalfa fork, four extra-length tines	$10-14
Fork, general-purpose, "Union Fork & Hoe"	$12-15
Vegetable fork, twelve tines, well-worn	$8-12
Coal or coke fork, "D"-shape short handle	$6-11
Spading fork, four triangular tines, squared back	$7-12

Gauges

Gauges both measured and marked objects, usually wood. Hard-used, many are made of metal, and the common types have a sliding collar on a long, calibrated rod. Early wood types can still turn up, a few maker-marked, of fine woods. More complex types move away from farm use and into the machinist's realm.

Values

Gauge, marking, brass thumbscrew, steel tip, 6 in. long $6-9

Gauge, panel; two-color wood, brass plate and lined fence, 21 in. long $35-45

Gauge, mortise and marking, brass thumbscrew, mahogany, 8 in. long $10-14

Gauge, panel, wooden screw, 20 in. long $14-17

Gauge, butt, steel, determines width and thickness, pocket-size $9-12

Gauge, mortise and marking, brass scale, fence and screw, rosewood rod, 10½ in. long $32-37

Gauge, butt, "Stanley #94", 3½ in. long $9-11

Gauge, mortise-roller, plated steel, 9 in. long $11-14

Gauge, plate, tap and drill, ¼ to ⅝ inches $19-24

Gauge, slitting, roller below handle, wedged blade, 18½ in. long $38-44

Gauge, slitting, one roller missing, 20 in. long $40-45

Gauge, stripping, "C. S. Osborne & Co.", 6 in. long ... $29-34

Gauge, stripping, cast iron handle, "Osborne/1876", 6 in. long $27-32

Gouges

These woodworking tools resemble chisels, but have a rounded lip or blade edge for making a grooved mark. Their "signatures" can be seen on many antiques, such as the liquid-channel that surrounds the compression chamber of an old farm press. Most gouge types do surface work on wood.

Early gouges can be all metal, with later types having a wood handle, perhaps with butt reinforcement in the form of con-stricted iron bands. A few types have a decorative brass sleeve at the handle and shaft junction.

Values

Gouge, straight, wood handle, 13 in. long............ $4-7

Marking gauges; top, all wood with wooden turn-screw, about 7 in. long. $6-8

Bottom, polished wood with brass turnscrew. $8-11

Above, Grain cradle, excellent condition, all parts present.
$50-60

Photo courtesy Fairfield
Antiques, Lancaster, Ohio

Bowl-making gouge, for hollowing wooden trenchers and bowls. This is the oldest metal farm tool illustrated in this book, being from the 1600's. Piece is 5 in. high, with rounded cutting edge.

$ - Museum grade

Grain scoops, ranging from large to small, with the last example about 9 in. long. Large (on left), wood and galvanized metal. $17-22

Medium-size, top right. $12-16

Small-size, bottom right; metal is old, heavy tin, age-darkened. $14-18

Gouge, wood handle, brass sleeve, faint maker's mark . . $7-10

Gouge, bent, "W. Butcher", cast steel, 11 in. long $14-16

Gouge, "WINCH" (Winchester), 14 in. long $19-24

Gouge, "Buck Bros.", cast steel, 15 in. long $12-14

Grain Cradles

Even into the early 1900's, grain crops were still being harvested with the cradle, one of the most spectacular of America's farm tools. It was popular because it let one man do two jobs. It cut the grain stalks, and deposited them to one side in a loose bundle, saving the gathering step.

Actually, the term "cradle" refers more to the attached wooden framework than the cutting instrument. The last was basically a scythe with an extra-long blade, and some extended up to 45 inches. The bend handle always had a lower handgrip, though the top might remain smooth. According to oldsters, it took years to become really proficient, laying out neat bundles of wheat, rye, oats or barley.

There were two main types of grain cradles, in addition to minor regional differences. The projecting wide-spaced tines or "fingers" varied in number, there being either three or four, with the latter more widely used. Four fingers allowed taller crops to be cut, but added weight to the work.

One problem with collecting these graceful-looking beauties is that often one or more of the fingers, or their spacer-supports or metal braces, are damaged or missing. Thus a grain cradle in good condition with all parts present is considered a prize find.

Values

Cradle, 39 in. blade, three fingers, average condition . . $45-55

Cradle, 41 in. blade, four fingers, average condition . . . $55-65

Hammers

Hammers are tools used to pound objects. On the farm this use ranged from tacks to chisels, horseshoe nails to barn spikes. By far the most common hammer was the wooden-handled carpenter-type, with striking surface and, opposite, the nail-claw or nail-puller. One interesting variety had a nail-holder on the side of the head, so that a nail could also be started without damage to fingers.

Special hammer shapes and weights were used at the forge, for tinning, even knocking snow from horses' hooves in winter. Wherever a swift and strong blow was required, there was a hammer designed for the purpose.

Values

Hammer, tack; lightweight, magnetic head, 9 in.
    handle $4-6

Hammer, carpenters'; hickory handle, curved nail-
    claws $5-8

Hammer, duo-headed, large and small striking sur-
    faces. $9-13

Hammer, farriers'; marked "Heller" $8-10

Hammer, file-maker's; rare, with good haft, 9 in. long . $85-95

Hammer, hoof; horseman's, nickel-plated, with pick. . . $28-35

Hammer, snowball; for cleaning horses' hooves, 10 in.
    long $24-29

Harness-Making and Repair

While the making of fine horse harness—either the heavy everyday work traces or the lighter, decorative harness for horse-and-buggy—was often done by a tradesman, the farmer yet had to possess some skill. Reins had to be replaced on short notice or a broken strap repaired.

Usually several rolls of bulk leather were kept handy for such work, perhaps in a barn tack-room. The leather was of tanned steer hide, dyed brown or black, and thick or thin as required.

The main instrument for harness-work was a sort of giant third hand, called by different names, like harness-clasp, harness-maker's bench, or stitching horse. All wood, it had a seat which faced a pair of wide, in-curved clamps. These closed tightly when a foot treadle was depressed. This held the leather firmly, allowing careful cutting or stitching.

Sharp-edged trimming knives were used to pare off excess leather, and types had stout steel blades and wooden handles.

Curved-blade knives were used for cut-outs and could handle the heaviest material.

A much-desired collector piece was a different cutting implement which vaguely resembled a pocket pistol. This was the leather gauge knife. The curved handle allowed good control with one hand.

A metal measuring rule was attached at right angles to the frame and had a sharp knife at the end. The user could thus get a standard-width cut and make long strips in a single motion.

Awls had long, needle-like projections with a sharp tip either straight or curved. Needles had an eye near the piercing tip for waxed harness thread. Holes were made with awls or the pliers-like leather punch, which had a revolving head with projecting tubes offering up different hole sizes. Holes were required either for rivets or buckle-tongues.

Riveting—fastening leather pieces together with metal (often copper) head and washer units which flatten and expand under pressure—was done in two ways. Pieces to be riveted could be placed on a metal surface and the rivet-head struck with a hammer.

Most farms, however, had a compact machine which operated on a fulcrum principle, applying great pressure on the rivet head as a lever was depressed.

To retain suppleness and strength, leather harness was oiled regularly with animal-fat derivatives, and sometimes darkened with lampblack. To retain strength, it was stored in a dry place, hung from harness pegs.

Values

Harness-making bench, clamp to top 38 in. high $65-90

Curved-blade harness-cutting knife $8-12

Straight-blade leather trimming knife $2-4

Leather-gauge knife, iron handle, cast, 4 in. long...... $30-40

Leather-gauge knife, cast-brass body, handle of exotic wood $65-80

Leather awl, 5 in. long............................ $1.5-3

Pliers-type leather punch, four tubes $7.5-13

Leather pliers, for holding and pulling material $4-6

Riveting machine, lever-type, 16 in. high $10-15

Tooling disc, for decorating leather, wood handle, iron
     wheel, 6 in. long $4-6.5
Harness buckle, cast brass, 2 in. long $4-6
Harness decoration, flower form, cast brass, 3 in. wide . $6-9

Harrows

Once the fields were plowed in fall or spring, additional
preparations were needed before planting could take place.
Even the rod breaking plow, which attempted to break up long
ridges of turned earth with projecting metal bars, couldn't do a
good job except in extremely dry or sandy soil.

The harrow was like a giant rake that cut into the first few
inches of soil. There were a number of types, all made to be
pulled and designed for breaking the earth into smaller pieces,
and leveling the surface of the field.

Three main kinds of factory-made harrows were widely used,
each purchased depending on the region of the country, type of
soil, and persistence of the salesman. Even earlier were stone-
weighted timbers pulled by horses, which did a rough and
general smoothing.

Toothed harrows came next, with heavy frames of hardwood
and teeth that resembled long railroad spikes without the head.
Pulled in angled rows, they were effective in good conditions
except that teeth were often lost encountering rocks and roots.
An improvement was the spring-tooth harrow, with large
curved teeth of tempered steel. They had clod-breaking
strength, with enough "give" not to be damaged.

Disc harrows, still used, had curved and round cymbol-sized
steel discs which were mounted on an axle beneath a heavy
metal frame. About 16 inches in diameter, the discs cut through
the soil with efficiency. Due to the special curvature, damp soil
did not normally cling to them.

Values

Harrows are another of that narrow category of farm implements
that have limited value to the average collector. They are mainly of
interest to the specialized collector or a museum. Some harrows are
valued so low that they are considered to be scrap metal, while others

are given away for the labor of moving and hauling. Too, some are purchased for actual use. While a single tooth from an early "spike" harrow may sell for several dollars at an antiques show, the entire rig may languish at a country auction. Collectors themselves will decide by degree of demand what will happen in areas such as this.

Hay Carriers

The hay carrier or hay "trolley" was the key piece of equipment for filling a loft or barn bay with loose hay. It was a compact cast-iron device about 16 inches long that was secured to the metal hay-track which ran just beneath the peak of a barn roof. Suspended from the track, and rolling beneath it, the trolley served several ingenious purposes.

It rode out to the end of the track and stopped beyond the open twin hay-doors. The hay-lift fork descended from it, to catch up several hundred pounds of loose hay. It was lifted straight up again by horse or tractor power on the rope, and when the load reached the track it shot along beneath the trolley at increased speed.

The carrier had greater velocity because it was now powered by a single rope instead of the double pulley-rope which doubled the momentum. Nudged to the correct place above the mow, the fork tines held below the carrier had the catches tripped by a lanyard, and the whole load dropped to the waiting pitchmen. The carrier was then released for a return and reload.

Carriers are not common collectibles because when baled hay came in—it made better economic sense and took up less space—the trolleys were no longer needed. Many were left to rust on the high track, out of reach of the scrap dealer and collector alike. They are used to some extent in "rural atmosphere" retail stores.

Values

Hay carrier, mounted with ropes on track section,
    educational display $60-75
Hay carrier, "Myer's OK Loader", good condition $25-35

Hay Forks and Rakes

After hay had been cut and left to dry for several days, it was ready for turning. In the horse-drawn era, hay was cut with the mower and sickle-bar before being gathered by one of two horse-drawn rakes. The first had a giant set of down-curved tines which pulled hay behind and released it in spaced bunches. Another, later, device, the side-delivery rake, whirled the dry hay into long, thin lines called "windrows".

Before such mechanization, hay forks and rakes did much of the work. Hay rakes gathered the grass, while forks were used both to turn hay so the underhay could dry and to pitch the long-stemmed grass onto the wagon. Soon after, forks were used to lift up bunches into the haymow and distribute it evenly. Damp hay was avoided, as it was prone to spontaneous combustion.

Hay rakes were simple affairs, being long wooden handles with a two-foot crosspiece at the working end which held a number of peg teeth. One problem with collecting them is that generally some teeth are missing, giving a certain, unbalanced appearance. Early examples were entirely handmade, while later types were factory products, sometimes with a metal or wire brace at handle junction. Some pioneer "hay drags" were quite large, and pulled by several persons.

Hay forks had crude beginnings, with the form derived from a plain forked tree branch. Some had but two projections and are considered primitive collectibles. Handles were always long, from 6 to 8 feet, either straight or slightly down-curved.

Hand-made types followed, and the number of tines ranged from the popular three to six or eight, and a few fancy types had spaced wooden reinforcement sections at the tine bases and metal tine tips for long wear.

Values

Hay fork, limb and forked branch, 6½ feet long	$18-25
Hay fork, handmade, wood, three peeled tines, 7 ft. long	$25-35
Hay fork, factory made, wood, three tines, 7 ft. long	$40-50
Hay fork, factory made, wood, six tines, 7½ ft. long	$60-75

Hay fork, factory made, three metal tines, 5½ ft. long . $8-12

Hay rake, peg teeth, three missing, 8 ft. handle $15-20

Hay rake, peg teeth all present, 8½ foot handle $25-35

Hay rake, primitive, 4 ft. wide, drag-type, 10 ft.
    handle $100-125

Hay Hooks

After the Great Depression of the early Thirties, hay balers (at first stationary, later mobile, tractor-pulled) began to be affordable on the average multi-purpose farm. These condensed work, in that one strong man, handling the twine or wire-wrapped bales, could perform with greater efficiency. The ever-present tool was they hay hook, for lifting and positioning the bales. Typically, only one hook was used per person, though some preferred two, one for each hand.

Hay hooks were simple and dependable. The one-hand base was wood or metal, the shaft and large, curved tip were iron or mild steel. (Sometimes these are offered by dealers as meat-hooks, but the latter have a smaller hook and longer handle, so that two men could lift a carcass.) Hay hooks range greatly in size, from 6 to 14 inches or so in length. Heavy-duty types may have been used in haying season by migrant "hands".

Typical hay hooks have a gracefully curved end, designed for easy sinking into a bale, whether of hay or straw. The angle is important, as the bale had to be detached with just a quick wrist movement. All were sturdy enough to take the weight of 75 and 100 pound bales, and one rarely sees a hook broken by flimsy construction.

Few persons understand that the hooks are mainly obsolete, no longer needed on the average farm.

Values

Hook, set in round wood handle, 9 in. long $4-6

Hook, wrought iron, stirrup-type handle with wood
    handpiece, 14 in. long $10-14

Hook, homemade from reinforcing rod, iron handle ... $2-3

Hook, wood handle, rectangular hook, 10 in. long $3-4

Hay Knives

Hay, the hybrid (high-bred) grasses of the cutting meadows, was the winter mainstay for cattle, sheep and horses. Before the arrival of the wire- or twine-wrapped machine-made bales, hay was stored as it came from the fields, in great loose piles. It was either put up in haystacks or went into the barn's upper level, the haymow.

The eventual weight of hay compressed the mass tightly, and when the stems and leaves needed to be removed for animal fed, it required special tools. One was the hooked hay tool, sometimes called a "crook", which pulled down armloads for the feedbins. Harpoon-like, of wood or metal, few are recognized for what they are.

To remove hay in volume, large knives were used. Early were the heavy-duty scythe-like blades attached to short, sturdy handles. They were designed to cut away easily managed bunches, and the blade is generally more impressive than hand-hold or footrests attached. All such types appear to be quite early.

Step-bladed or sawtooth versions were popular in the Mid-west, these being curved and with an end and a side-handle.

Still another type is the chisel-bladed hay tool, with straight handle and narrow blade with three to five sharp triangular edges. It was designed to thrust sharply into tough hay. In the present era of machine-produced very large hay rolls, tractor lifted, the hay knife is no longer in use.

Values

Hay tool, hand-carved wood, hooked end, early, 6 ft.
long $25-32.5
Hay tool, iron, factory-made, hooked end, oblong
enclosed end handle, 5 ft. long $8-12
Hay knife, iron scythe-size blade, short wood handle . . . $20-30
Hay knife, sawtooth edge, two wooden handles $10-15
Hay knife, spade-like blade, wood handle, serrated
edge. $7-10
Hay knife, spade edge serrated, five extensions $10-13

Haymow Forks
(See also Hay Carriers)

The large forks used to offload loose wagon hay with the hay carrier were typically the two-pronged and double-harpoon type. This design was widely used for all varieties of hay, being lightweight (less than 20 pounds) and practical.

Other mow forks were used for heavy or long-stem hay like timothy. A single-shaft harpoon fork, robust in construction, had two catch-extensions near the tip. It was good for accurate placement of smaller loads in the mow.

Another type resembled a giant claw, and was called a "grapple" fork. Some varieties had opposing sets of two tines and were used for average haying conditions, as was the double-harpoon configuration.

An offshoot was a grapple fork that had three tines per side, for a total of six. Massive, with weight over 50 pounds, these forks were used for short-stem hay like alfalfa, and for loose grain-straws.

Values

Double-pronged harpoon fork, about 29 in. high, with
  some original red paint intact $22-26
Single-prong harpoon fork, heavy, somewhat scarce . . . $25-30
Grapple fork, four-tined $30-40
Grapple fork, six-tined, good condition $40-50

Hinges

In the days of handmade objects ranging from boxes to doors to barns, iron hinges were usually obtained separately. Factory-made examples abound, whether they secured a blanket-chest lid or the door to an in-barn granary. Some factory products—almost always made in matched pairs, as were hand-wrought examples—are in demand by collectors, especially if found with maker's mark and/or date.

More sought-after are the older handwrought hinges from the blacksmith's forge, those often being large enough to secure the quarter-ton hay doors to the second storey of a barn. The hinges

Hay loader for elevating hay from loose windrows, pulled behind hay wagon.

Photo courtesy of Hubbell Trading Post National Historic Site, Ganado, Arizona.

Below, Hay hook, wrought-iron, 9 in. long.
$6-9

Left
Leather-working tools; top, leather marker 7 in. long, brass ferrule, cogged steel wheel. $7-9
Bottom, leather-punch, brass base below hollow pins, 9 in. long. Marked "Lodi /Schollhorn Co.".
$8-10

Harness brass for connecting and decorating horse-harness, with examples to left and right 3 in. wide.

Top	$4-7
Bottom	$2-3
Left and right, each	$7-10

Disc harrow with front and rear disc sets.
Photo courtesy of Hubbell Trading Post National Historic Site, Ganado, Arizona

Side-delivery rake for rolling hay windrows, here with early rubber tires, Maker-symbol is a joined "IH" inside a large "C".
Photo courtesy of Hubbell Trading Post National Historic Site, Ganado, Arizona

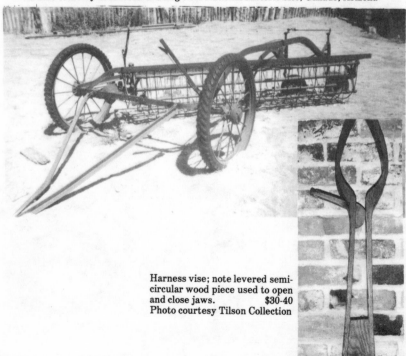

Harness vise; note levered semi-circular wood piece used to open and close jaws. $30-40
Photo courtesy Tilson Collection

were carefully made for strength and shaped for beauty. Collectors have names for attractive types, based on the resemblance to certain common animals or objects.

Single units are acceptable if the hinge is old, forged, in good condition and of an unusual type. All are considered to be homespun art of the first order.

Values

Birds-head ends, pair, plain straps, 18¼ in. long $70-90

Double-arm ends, pair, 23¼ in. long $50-70

Double swan's head, a pair, 23 in. long $55-75

Hinges, pair, plain, 32 in. long $45-65

Pitchfork-shaped, pair, 8½ in. long $45-65

Ram's horn, pair, 9½ in. long $40-60

Snake's head, single hinge, 15 in. long $40-60

T-shape, single hinge, swordfish ends, 36 in. long $60-80

Turnip-shape, single hinge, 6¾ in. long $12.5-20

Hired Man's Bed

Even with the aid of older children, a farmer did not always have enough help, especially at harvest time. The last good days of fall were worked from dawn to dusk, even into the night under the "harvest moon" of the autumnal equinox.

It was a common practice to obtain additional adult workers, the "hired hand" or "hired man". There was little problem with setting an extra place or two at the table, more difficult finding a place for the night. This resulted in what is known today as the "hired man's bed", sleeping furniture that was constructed low enough so that it could be slid under a normal-height bed.

Actually, hired hands did not usually sleep in the farmer's own bedroom, but were often put up in a spare room, an outbuilding, over the springhouse or in the loft of a summer kitchen. The main benefit of a bed of this type was that two stored easily in the space for one. Children or unexpected company undoubtedly used it more often than seasonal laborers.

Values

Bed, hired man's size, walnut, refinished............. $140-160
Bed, hired man's, rope support, turned legs, size
    28x65x16 in. high $65-75
Bed, hired man's, turned posts, old red paint, orig. rails,
    size 30x77½x23 in. high $360-400
Bed, hired man's, cherry, rope supports, old paint $95-115

Hoes

The first American farm hoes were copied from, even made
for, the trader with Indians. The hoe-heads were hand-forged,
had a large round or oval eye, and a somewhat unwieldy blade.
Nonetheless, they were superior to other materials and better
than weeding by hand.

Hoes have entered rural folklore, with the words and tunes
about a young man that wouldn't hoe corn, to hoe-downs, to
things being a tough row to hoe. That aside, hoes were blades
on long handles, made for small-scale agriculture. Hoe-heads
were shaped differently for selected tasks; prime types are,
garden, triangular, "scuffle", weeding , grub and ginseng.

Garden hoes need little comment, and are rounded or
squarish. Triangular or pointed hoe-heads were used for deep
and careful weeding, while the rectangular, double-bladed
scuffle hoe was used for root-cutting near the surface. Weeding
hoes had a blade and opposing this, a two-tined projection.

Grub (grubbing) or digging hoes had a large and long blade
that went deep, and could be used for taking up root crops. The
ginseng hoe was a lightweight tool for taking to the woods and
recovering wild "sang" or ginseng roots, this forked part of
reputed medicinal value.

Values

Hoe-head, trade era, ca. 1760, hand-forged, round eye. 60-70
Garden hoe, old, worn handle, no maker's mark $3-5
Triangular-head hoe, 4½ ft. handle, good condition ... $6-8
Scuffle-head hoe, no handle, unusual, old............ $9-12
Weeding-head hoe, handled, maker's stamp.......... $6-9

Grub-head hoe, handled, unused condition $12-16
Ginseng hoe, handled, 34 in. long, light double head . . . $25-30

Horn Items
(See also Powderhorns)

The horns of domestic cattle, while integral to the animals, had a more permanent use to owners. The farmer might save a pair to commemorate a prized bull, mounting them in the barn or on an outbuilding. But the real value came when horns served a purpose.

In Colonial times, a horn tightly sealed with pitch or beeswax sometimes served as a salt container, protecting the contents from moisture. Much more widely used were the grease horns, repositories for bear grease, bacon drippings or tallow. These were necessary for greasing wagon wheel hubs to reduce friction and damage.

Grease horns tend to be found in pairs, perhaps one for each side of the vehicle. The small end is left intact and the large end closed with a wooden block, with center hole open for the dauber stick. Many examples are iron-bound at the large end for extra strength. Most of these horns have a suspension strap or chain.

Hunting horns had little done to them besides the usual scraping and polishing. Both large and small ends were left open; they were used to sound notes for directing hunting dogs or for signaling to other hunters. Hunting horns are usually fairly large and some have wood or metal mouthpieces.

Values

Horn salt container, hinged opening, wood end, early
    1700's $45-60
Grease horn, single example, 13 in. long, with iron
    chain $15-25
Grease horns, matched pair, large ends iron-bound,
    12 in. long, chain suspensions $30-40
Hunting horn, 14 in. long, carved-wood mouthpiece . . . $25-35
Hunting horns, matched pair, 13 in. long, suspension
    cords of heavy cotton string $30-40

Hay baler, stationary, but sometimes pulled in conjunction with hay loader and extra workers. Maker stencil on side: "Made by Kansas City Hay Co.".

Photo courtesy of Hubbell Trading Post National Historic Site, Ganado, Arizona

Below
Grease horns, matched pair, wood-plug ends with dauber hole, iron-strapped and chain-linked, about 13 in. long. $30-40

Right, Digging hoes, small blades, rounded tip. Example on right was fitted with an early axe handle. Left, straight handle 3 ft. long. $5-7
Right, with axe-handle $6-8

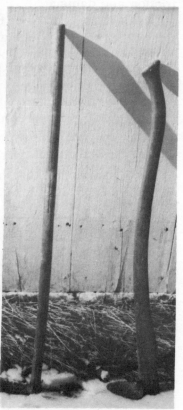

Above, hunting horns (**not** powderhorns) used to call and control dogs while hunting. Open-ended, matched pair, each about 11 in. long. Pair
$25-35

Horse Tethers

When horses were the primary means of rapid land tanspor-tation, in all of the 18th and 19th centuries, the horse tether was of great worth. Like taking the car keys or putting on the parking brake in a modern automobile, it assured the owners that his method of conveyance remained where he left it. Horse tethers are of two broad classes, portable and permanent.

Portable tethers are of the "weight" type, heavy cast-iron masses with a top tether-ring. These were carried in the buggy or carriage, and deployed when the destination was reached. The weight was placed on the ground before the animals and they were fastened to it with a short cord.

Most such farm weights are circular and tapered, with the broad flat side placed on the ground. The near-universal weight is 20 pounds, more than a horse could easily manage. Basic and practical, only a few weights were cast in pleasing, "fancy" forms such as a beehive.

A horse and rider could not carry such a weight so another device was used in early days. This was the sticking tether, of iron, with a sharp tip and end with two curled loops for reins. Locally made by blacksmiths, and with regional variations, the tip was pushed into a handy log or tree. These are sometimes called "reins keepers".

Permanent tethers, in front or beside the farmhouse, included the single ring in stone or a plain tether-post. This sometimes is in the form of a cast horsehead on a pole. A Black figure of a jockey, often painted, served as a surrogate footman in some areas.

Values

Horse tether, 20 lb., plain, rusted but hitching ring
    intact $20-30

Horse tether, beehive configuration, unusual $40-55

Paperweight, advertising, in shape of a miniature horse
    tether, cast iron $15-20

Reins keeper, blacksmith-made, sticking type, 7 in.
    long $25-30

Plain post iron ground tether, 3 ft. high $8-14

Horse-head iron ground tether, head 9 in. high $50-75
Jockey-type tether, faded paint, original casting $90-125

Ice Skates

Skating was always a pleasurable part of winter on farm ponds and lakes. This reached a sort of mania level in the late 1800's, and almost anyone skated, or tried to.

Old skates are still found in some numbers, and collectors seek very early types with up-curled runner tips, screw-to-shoe models, and those with unusual shapes or by long-defunct makers. Matched pairs, of course, are the most highly admired.

Values

Ice skates, pair, wood with steel runners, brass heel and
    toe plates $24-30
Ice skates, pair, early wood and iron with upturned blade
    tips $42.5-50
Ice skates, clip-on, nickel-plated, steel, boys' size $12-15
Ice skates, pair, wrought iron runners, "C. W. Werth/
    Warranted", 13½ in. long $50-60
Ice skates, pair, iron, worn leather straps, 11¾ in.
    long $30-40
Skater's lamp, tin, "Perko Wonder Junior", 6½ in.
    high $30-40
Skater's lamp, all brass, steel bail, 10 in. high $50-60
Skate sharpener, stone in metal holder, 4 in. long $7-9

Ice Tools

Extensive and well-to-do farms of whatever Midwestern location often had a low-slung building, partly underground, put up on or near the shore of a pond or lake. This was the ice-house, designed to keep ice even into the hottest summer months. It was insulated with below-ground siting, roof overhangs, double walls or roofs, plus tight layers of hay or sawdust. Windowless, it was made to keep heat out and cold in.

Ice "harvesting" was done in the dead of winter, when the ice was of extremely low temperature, and it was safe to take men and equipment out on the frozen layer. Ice at least 16-18 inches thick was preferred, for the large blocks made ice-taking efficient.

A number of special tools are associated with this frigid task, beginning with the ice-auger which made the first clean hole. Horse-drawn ice-scorers were used in commercial ice operations, but the small farm did not have such refinements. Instead, cutting lines were made by "eye", and rectangular blocks were cut to a manageable size.

Once the first hole had been augered or chopped, the ice-saw was used. This was a unique implement, suitable for no other task, designed only for cutting thick ice. About 4 to 5 feet long, of iron, it had a square tip, large teeth, and a wooden crosspiece handle mounted at right angles to the blade flats. Used with both hands, the worker backed along the desired linecut, sawing as he moved. Eventually the cutting area resembled an elongated giant checkerboard.

Chopping saw-holes and maneuvering blocks in the now-open water required two additional tools, the ice-hatchet and ice-axe. As with most tool descriptions, the former was a smaller version of the latter. Each had a wood handle, up to 18 inches for the hatchet, a yard for the axe. Each had a long, thin chopping blade, and, on the poll area, a slightly curved hook. Either could be used from solid ice for chopping or maneuvering the blocks. As with log-handling, a pike was also used.

Lighter pikes were made, and had long handles, with either a straight or hooked tip. Tongs were used to lift and slide the ice blocks into the shed interior, usually up a slanted ramp. The tongs exist in one and two-handled versions, but all are heavily constructed.

Such vanished tools remind us that summer coolness did not come from a plug in the wall, but a place on a lake.

Values

Auger, opposing wood handles, 4 ft. long $40-50
Saw, hand-forged, rectangular heavy blade, wood
    handles $55-65
Saw, steel blade, factory-made, 5 ft. long $40-50

Hatchet, early, narrow blade, curved tip	$35-45
Hatchet, factory-made, narrow steel head, curved tip	$30-40
Axe, head 14 in. long, curved poll-tip, steel, handled	$40-55
Axe, steel head, curved tip, broken handle	$30-40
Ice-pike, farm-made, single straight tip, 7½ ft. long	$15-20
Tongs, two-handled, large, wrought iron, 26 in. long	$35-45
Ice auger, hand-forged, 40 in. long	$35-40
Ice axe, "Underhill Edge & Tool Co.", lightweight	$40-55
Ice chisel, with orig. wood handle, blade 17 in. long	$35-40

Implement Seats

A small but devoted coterie of collectors are converging on cast-iron implement seats. These were once attached to horse-pulled equipment from mowers to seeders. Many a man (the writer included) can recall their unrelenting solidarity and the violent jerks and lurches they dispensed.

For all the interest, their period of popularity lasted only from the late 1860's until around 1900. They came into being because workers wanted something better than walking beside or behind the equipment. The cast-iron seat let them ride, but they were eventually replaced by sheet-steel seats, cheaper to make and easier to sit on.

Cast-iron seats are characterized by bolt-on bottoms, ventilated (oval holes) bottoms and sides, and a rounded rim. Most are severely practical, even plain, but a few have well-done designs. A number have maker's marks worked into the seat structure. These are more valuable than plain specimens, as they may be better-made, more attractive, and can at least be researched as to maker.

These implement seats are in short supply for several reasons. One is the relatively short period of manufacture, just over 30 years. The other is the great patriotic WW-II scrap-metal drives. But, such interesting "farm supports" turn up with some regularity, and at least one seat adds credibility to the general farm-related collection.

Values

| Plain cast-iron seat, rusted, unmarked | $10-15 |

Ornamental seat, not maker-marked, star designs $35-50
Ornate seat, oiled, figural decorations, "Ideal" $75-100

Indian Artifacts

Indian artifacts are very much a farm-related item. Just as farms were sited in good land for crops, so prehistoric American Indians selected pleasing surroundings for their villages. In the course of farming activities, most farmers picked up the scattering of artifacts, the cultural debris, as they surfaced. It was a rare farmer that did not have a collection of some size, likely stored in the favorite receptacle, an old wooden cigar box.

The largest artifact group is known as "projectile points". Since only late and small types are true arrowheads, the points exist in probably 500 different sub-styles in the Continental U.S. Most were chipped from cypto-crystalline materials like flint and obsidian. Though many are somewhat damaged from years of harsh contact with agricultural equipment, they still are picked up on farms and ranches.

The second large group are the hardstone tools, these pecked and ground into the final shape and then sometimes polished. These would include chopping axes, full-, three-quarter- and half-grooved classes, and the later ungrooved celts. Also included would be grain-grinding pestles, hollowed-out mortars, and grooved hammerstones.

The third class of Amerind artifacts likely to come under the hammer at a typical farm auction would be the slate objects. This is not the thin roofing slate, but colorfully banded material deposited long ago by glacial activity.

Many fine objects were created, such as the single-hole pendant, the two-or-more holed gorget, and the esoteric Atl-atl weights, used to decorate the lance-throwing stick.

One of the greatest farm finds would be a birdstone, named because of a certain resemblance to the profile of a bird setting a nest. The most common form of "birds" is made of slate in a number of styles, elongated to bust-type, and collectors vie for good pieces at $1200 to $1500 or more.

More rare still is the hardstone "bird", made from granite or porphyry. A few years ago, one was found in the bottom of a box of mediocre tools and parts, auctioned off at two dollars.

Above, **wooden** implements; top, "stomper" about 3 ft. long, large wooden head, used probably for making sauerkraut. $7-10

Bottom, large stirrer about 4 ft. long, used for mixing in kettles or other large container, perhaps for cheese-making. $9-12

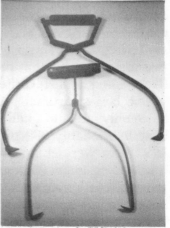

Indian artifacts, left item 5 in. high.
  Left, Hopewellian celt, high polish $25-30
  Center, full-grooved axe, well-polished, battered edge $40-50
  Right, celt, little polish $18-25

Ice-tongs, one-hand size, cast-iron with wooden handles. Top, with 12 in. arms. $9-12

Bottom, of heavy iron wire. $7-10

The lucky buyer had a superb Amerind object, photos of which appeared in several amateur archeological publications. Hardstone "birds" start at about $3000.

The highest category of Amerind artifacts likely to come from Midwestern farms are the effigy pipes, made in late-prehistoric Woodland times by certain mound-building Indians. True fine art, they often depict animals. Such ultimate objects are considered to have no upper values to limit them, being of museum grade.

Values

Point, grey glossy flint, Kentucky, 1¼ in. long $2-3

Point, side-notched, brownish obsidian, Washington,
    2 in. long $12-15

Point, base-notched, New York, 1½ in. long $3-4

Point, pentagonal, Flintridge, Pennsylvania, 2¼ in.
 long $16-20
Point, Ft. Ancient triangular, Ohio, serrated, 1¼ in.
 long $4-6
Blade, triangular, Texas, corner-notch, 3⅞ in. long $28-32
Blade, Adena, tan, Illinois, 4⅞ in. long $50-60
Blade, black and tan, Missouri, 5 in. long $125-140
Celt, polished, Arkansas, 4¾ in. high $30-35
Celt, grey, Michigan, 6 in. high $25-30
Celt, porphyry, Ohio, overall polish, perfect, 8 in.
 high $100-120
Axe, half-groove, tan sandstone, Iowa, 5 in. high $25-30
Axe, full-groove, black, polished, Indiana, 4½ in.
 high $30-40
Axe, three-quarter groove, Tennessee, 7½ in. high $60-70
Axe, three-quarter groove, Wisconsin, fluted, 9¼ in.
 high $150-175
Hammerstone, Illinois, three-quarter grooved, 3 in.
 high $12-15
Pendant, banded slate, Pennsylvania, 1½x3 inches $25-30
Pendant, banded slate, Missouri, 1⅝x4¼ inches $40-50
Pendant, banded slate, keyhole type, 4½ in. high $125-150
Gorget, banded slate, Alabama, 2x5 inches $45-55
Gorget, banded slate, Illinois, broken at hole, 2x3⅛
 inches $9-12
Gorget, hardstone, polished, from unknown state,
 1¾x5¼ inches $200-225

Journals

Whether the items are referred to as journals, ledgers or
diaries, the meaning is only slightly different for this farm
record-book. An old farm journal is an on-the-spot momento of
life at that time.

Ledgers tend to be closer to accounts books, with products or
services rendered, and cash/credit notations. A diary, as the
name suggests, is a daily series of comments, generally what
happened in a certain year. While a diary may be a day-log of
personal experiences and observations, the journal may range
further and include anything from birth dates, crop yields, the
weather or social contacts.

Farm journals are not especially common, and those a century and more in age can be considered rare. The early examples have leather covers, either tanned or rawhide, and the interior sheets are handmade paper, often yellow with age.

Stamp-impressed designs sometimes decorated early covers, while late-1800's journals might be factory made and have ruled sheets. A number of the oldest journals had a thong wrap for fastening the covers together.

Perhaps the best-known very early journal was that of the boy, Noah Blake. Written in 1805, the work was discovered and published a few years ago by the noted writer-illustrator Eric Sloane. This provides an excellent account of rural New England life in the dawn of the 19th Century.

Another journal, owned by the writer (see photograph) has a number of entries, the first in German and dated 20 May 1833. Early writings were done by quill pen in a brown ink. Folk remedies are given, and one, for horses, contained (in the original spelling) the following ingredients: "Sassafrac bark, 2 oz, Lichrich root, 2 oz—One spoonful twist a day". Another included "1 hnful grene rye, 1 hnful fishworms".

Such journals from bygone days give a valid glimpse into the farming past. The minus is that few were faithfully kept and fewer still survive in good condition. The plus is that early journals should become treasured collector items for the many persons who now collect farm items.

Values

The quality of farm journals varies greatly, and depends on factors like condition, age, number of entries and the extent of detailed observations, and so forth. Furthermore, a local farm journal (say from the same county) will bring more at auction than an out-of-state example. Though difficult to evaluate, most are in this range: $20-45

Jugs
(See also Crocks)

Stoneware jugs differ from crocks, in that jugs were designed to hold only liquids. Whatever the body shape and whatever handles are present, the neck was narrow, as was the lip and pouring spout, if present.

Such jugs were used on the farm for everything from water-carriers for thirsty field-hands to liquor and honey containers. Molasses, boiled-down sorghum juice, was kept on most farms as a sweetening food ingredient or for medicinal purposes. Jugs also held various folk-healing concoctions, most so awful-tasting that recipients knew the brew had to be good for them.

Early farm ledger or journal, rawhide cover, thong fastener, open to earliest date, 1833. $25-30

Stoneware jug, one-gal. size, brown glazed top over white, air-hole and spout in top. $15-20

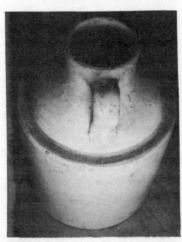
Wide-mouthed jug, stoneware, white glaze, 12 in. high. $12-15

Values

Jug, four-gal., thistle-and-swirl design, gray $175-210
Jug, cobalt floral-and-swirl design, 11½ in. high $125-150
Jug, "Charlestown", one-gal., beige, 13 in. high $85-100
Jug, gray, salt-glazed, cobalt feather design, 12 in.
  high $85-100
Jug, three-gal., cobalt blue feather design, 15 in. high . . $65-85
Jug, impressed "H. & C. Nash/Utica", 10½ in. high . . . $80-95
Jug, five-gal., "Weading & Belding/Brantford, Ohio" . . $65-80
Jug, signed "Bennington, Vt.", ovoid, 10 in. high $110-130
Jug, three-gal., "A. H. Wheeler Co./Boston, Mass.",
  30 in. high $175-195
Jug, "Whites Utica", bird in cobalt slip, 22 in. high $155-175
Jug., two-gal., stylized floral design, 25 in. high $70-95
Jug, "M. Hanssling/Newark, N.J.", 14½ in. high $50-60

Kettles

All farms had large kettles for outdoor use, and this ranged from small-scale maple sugaring to butchering. The kettles were also used to make communal stew for gatherings, apple butter making, even clothes washing and dyeing. In the days of large families and much food demand, a kettle was always on standby alert.

Kettles—large angled-bottomed or round-bottomed affairs, holding from 20 to 40 or so gallons—had a sturdy rim and heavy handle, usually of wrought iron. Cast iron kettles are of several types, footed and non-footed. The last have three or four cast projections to raise them several inches from the ground or coals. It could also be hung over an open fire by the cast or wrought bail.

More desirable and interesting are the numerous copper handmade kettles, lapped and dovetailed with brass at the joins. Examining a copper kettle closely is an educational experience, from method of wrap-join to the iron inner lip, bail, and handle ear rivets.

Both iron and copper kettles sometimes had wrought iron stands, usually tripod-type, and the stands are more scarce than the kettles themselves. The circular stand portion is usually an iron strap, the legs strap or rounded. Unholed copper kettles in good condition bring a premium, as do cast iron kettles without chips or cracks.

It is well to note that many of the larger spun-brass buckets, some with maker "signature" on the outside bottom, were also used as small kettles. Many in fact have a hanging bail and evidence firemarks on the bottom.

Values

Kettle, copper, double row of brass dovetailing, 25¾ in.
diameter $395-450

Kettle, hand-forged iron handles, copper rivets, 7½ in.
high $295-350

Kettle stand, iron, openwork top, 13 in. high $55-70

Kettle, cast iron, on three legs, iron bail, 16 in.
diameter $80-95

Kettle, copper, several see-through pinpoint small holes,
handle rusted, 17 in. high $85-95

Kettle, cast iron, long crack, rounded bottom, good bail,
14 in. high $25-35

Kettle, copper, on iron stand, prob. matched, all in good
condition, kettle 15½ in. high $375-450

Kettle, bell-metal brass, wrought rattail handle, brass
ears, 10 in. top diameter, 10¾ in. high $175-200

Kettle, cast iron, 5 in. diameter, maker-marked, old ... $22-30

Kettle, copper, "tulip" handles, attached link chain for
suspension, brass dovetailing, 9½ in. diameter.... $160-185

Kettle tilter, wrought iron, 18th Century, graceful $265-300

Knives

From frontier explorers to pioneer farmer to Victorian rural gentleman, the knife was of key importance. Formerly carried as a large blade at the belt, sheathed in leather, the knife became more refined as social conditions modified.

The large, lethal blade was reduced to one that was both more manageable and gave better protection to the owner's hand and fingers. This was the ubiquitous pocket-knife of many sizes, with one or more special-purpose blades that folded edge-down into the handle.

Hunting knife, 7 in. blade, brass guard, bone handle,
"Sheffield/EBRO". $45-60
Photo courtesy Fairfield Antiques, Lancaster, Ohio

Barlow and Case were great names, and there were many
others. Such knives were invaluable on the farm, being useful
for cutting binder and bale twine, feed-sack ties, or grafting
saplings.

Whiling away time alone before the TV-era was likely spent
employing the knife in imaginative ways, or whittling. This
was making objects from raw wood, controlled by the eye and
skill of the maker. Spiles (for tapping maple trees) might
emerge, or a willow-branch whistle, even the curious creations
known as "whimseys". The maker might produce anything
from solid-wood chains to animal figures to cunningly carved
walking sticks or cane-heads.

The Victorian rural gent, distancing such vulgar diversions,
still carried the diminutive penknife, with handle of mother-of-
pearl or similar exotic material. Often carried on a watch or
keychain, such knives were used to sharpen the points or nibs of
quill pens.

Values

Case pocket-knife, green bone handle, folded 6¹⁄₁₆ in.
    long $50-80
Winchester pocket-knife, celluloid handle, 4⅛ in.
    folded $60-90
"Keen Kutter Kattle Knife", three blades $40-60
Schrade "Stockman", 3⁵⁄₁₆ in. folded $12-15

Lanterns

The need for farm lighting during the dark hours gave rise to
many different light-producing devices. While lamps were

mainly used in the farmhouse, lanterns could generally be used both indoors and outside, with provision for carrying and protection from the wind. Due to the danger of fire, farm lanterns have a protected flame. The word derives from old terms meaning "shining" and "horn", and very old lanterns had thin horn windows.

Fuel to fire the lanterns came from many sources. Grease and tallow supplied early devices, followed by sperm-whale oil, coal oil and kerosene, still later by "white gas" or unleaded gasoline. Along the way, other examples burned compressed natural gas, even carbide gas. Electricity, of course, eventually prevailed with the advent of wet and dry batteries.

Whatever the period or type, lanterns had to be safe, portable, economical and dependable. Assisting a calving cow or a ewe laboring under the bulk of triplets, the farmer literally could not work in the dark. In an emergency, household lamps could be pressed into service, but lamps and lanterns had different designs and uses.

Values

Barn lantern, tin, blown chimney, brass burner, "L"-shape foot for beam straddling, iron loop holder $45-65

Candle lantern, seven glass panels, top with handle; vent hinged for access, 11½ in. high $110-125

Candle lantern, glass on three sides plus tin panel; attached match-holder, lantern 10½ in. high $40-50

Candle lantern with carrying case, mica front, tin sides and back; whale oil burner fits candle socket, 10 in. high $75-90

Candle lantern, tin, pierced design, top ring, 12 in. high $80-100

Candle lantern, tin, three glass panes, 12½ in. high ... $90-125

Hand lantern, tin, folding; has two mica windows, size unfolded, 3⅛x3¾x5½ inches $85-100

Kerosene lantern, bullseye lens, 13 in. high $17.5-25

Kerosene lantern, clear blown chimney, tin base, 14½ in. high.................................... $24-30

Lantern, iron and tin, clear glass globe, decorative pattern of holes, 11 in. high.................... $130-150

Lantern, tin, Paul Revere type, 14 in. high $65-85

Lantern, early pierced candleholder type, 16½ in.
high $115-135

Lantern, tin, wall-mounted, removable vent finial,
28½ in. high $90-115

Lantern, sheet iron with wrought fittings, red glass,
brass oil burner and font, 15 in. high $55-75

Lantern, tin, orig. ring and loop, gray-black tin, hand-
pierced decoration; fine condition, ca. 1820, 16 in.
high $150-180

Lantern, "lighthouse" type, tin, wide armstrap, light-
house-shaped globe, mold-blown, rows of stars at
top vent, whale oil reservoir, ca. 1830, 17 in. high . $195-225

Latches And Locks

The integrity of farm buildings was always important, from the earliest interior door-bar activated by the exterior latch-string to the latest in all-metal locks. Latches were fastening devices, often operated by a simple catch and hook or shot-bolt arrangement.

Locks were more complicated, requiring a key—either a device or special knowledge—to open. The name "padlock" derived as a protection against "footpads", sneaks in the night.

The first American farm latches were all-wood, simply a dropping or sliding bar on the inside of a door. Soon metal fittings, hand-forged of iron, bolstered the bars and they eventually became all iron or steel, then factory-made. The hasp is a specialized latch, usually of metal, with a tongue that drops into a catcher of the same material.

Locks, more intricate, were designed to prevent entry into a building or protect the contents of a container. There are quite old locks, all of wood, put on certain doors (smokehouse) activated by wooden keys that have disappeared with time. In general, the more valued the contents, the bigger and better the lock.

Collectors look for handmade locks of whatever types; those with a "pedigree" and record-proofed to have once been part of an historic building are a top prize. Hand-forged latches are good, also those with an ingenious design. The factory-product locks should be in good condition, with the original key, hopefully with brass or bronze parts or body and the maker's mark.

Values

Latch, file slider, U-bold catches, farm-made, early ... $4-6

Hasp/latch, barn door, hand-forged, lightly pitted,
  12 in. long $25-32

Thumb latch, iron, early, cast handle, all hardware,
  8¼ in. high $27.5-32.5

Thumb latch, iron, early, hardware, 7 in. high $25-30

Padlock, "Yale/Y. & T.", all metal, bronze case, self-
  closing, no key $10-12

Key, iron, early, 4½ in. long $2-3

Key, brass, folding, for outbuilding, late 1800's $12-15

Lock, iron and brass, keyhole cover with engraved
  eagle, 2¼x2⅞ inches $75-90

Padlock, "Empire 6-Lever", brass, compact $11-15

Lock, iron, heart-shaped with heavy solid brass keyhole
  and plate, 3½ in. long $23-27

Padlock, hand-made, cylinder and hook type, 9½ in.
  long $28-33

Padlock, iron, attached wrought iron eyebolt, 3½ in.
  high $15-17

Lock, iron, 4¼ in. high, good condition $10-13

Padlock, "Slaymaker/Shark", brass body............ $9-15

Box lock, iron, early, brass knobs, key and brass-
  trimmed keeper, brass label with eagle, 4⅜x5
  inches $77.5-90

Lock, elbow, wrought iron, pitted, with keeper, 5x6¼
  inches $50-60

Lock, wrought iron, with key and snakehead keeper,
  5¾x6 inches $65-85

Lock, wrought iron, old, 5½x10¼ inches $15-22

Padlock, with key and chain, "Armory Eight Lever",
  3½ in. high................................ $7.5-10

Doorlatch, iron, polished brass knobs and keeper,
  3¾x6½ inches $55-70

Thumblatch, wrought-iron, tulip top, pivot pin missing,
  10 in. high................................. $160-190

Levels

Instruments used to gauge and establish horizontal planes,
levels are used to determine what's level or exactly horizontal.

Made of metal, wood, or wood with metal end-caps and gauge fittings, most levels are relatively long and narrow.

Levelness is measured by liquid-filled glass tubes with an air bubble which can be centered between marks midway between the tube ends. They are also called "spirit" levels because the liquid, often colored so the bubble can be easily seen, contains a high percentage of alcohol. This serves as an anti-freeze agent so levels can be used in cold weather. Otherwise, too, the liquid would freeze and break the tubes.

Levels range from 3 inches to over 2 feet in length, with metal examples generally smaller than wood or brass-strapped types. With the appropriate extra tube, many levels can also be used to determine vertical angles, but only because the second tube is still positioned in a horizontal plane in the level.

Values

Level, "Henry Disston & Sons", brass ends, good glass, 28½ in. long $35-40

Level, "O.V.B. #100/Pat. 5-8-06", brass ends, rosewood, vertical bubble missing, 3x27½ inches $45-55

Level, "Stanley Rule & Level Co.", good glass, brass ends, "Pat. 11-11-1862", 27 in. long $42-48

Level, "Stanley #36/Pat. 6-23-96", perfect, 9 in. long .. $35-40

Level, "Stanley #37", perfect, 2x6 inches $42-48

Level, "Stratton", brassbound, two bubbles with wire centers, 8½ in. long $65-75

Level, "Stratton Bros./Pat. 7-16-1872", brassbound, 6¾ in. long $80-95

Log Cabins

Mention is made elsewhere in these pages of the famous log cabin, as being one of the first farm buildings, and later occasional use as an out-kitchen or loom-house. Being solidly constructed and treated down through the years with a certain awe and respect, it is not surprising that examples survive and are sometimes offered for sale.

Here is a brief listing of some recent examples; while these values may seem high, small outbuildings such as smoke-houses

Level; good wood, brass and alcohol bubble-tube. "Stanley Rule & Level Co./ New Britain, Conn. U.S.A./Patd 11-3-08", 24 in. long. $22-26

Wedges; left, for splitting wood, 5 in. high. $3-4

Right, wedge for stonework, from an old farm quarry. $5-6

Chain gear; background, section of logging chain with one-inch links, several feet long. $4-5

Foreground, chain connecting links, both well-worn, used with chains or equipment, 3 in. wide. Each . . . $2-3

Brass-barreled pump, for pressurizing early gasoline lamp; wood handle, 11 in. long. $10-13

Photo courtesy Fairfield Antiques, Lancaster, Ohio

can sell for as little as one hundred dollars. A major factor, of course, is that the buyer usually must arrange, and pay for, transportation to another site. In the case of log cabins, this usually also means a step-by-step reconstruction at the new location. Listings below are of cabins well over one hundred years old, authentic original dwellings.

Values

Log cabin, 14x18 ft., with loft; good condition,
original wooden shingles in place beneath tin, oak
logs .... $4000/"Or
best offer"

Log cabin, one-and-a-half storey, 12x16 feet, all wood
good, partially taken down and parts keyed $3800

Log cabin, two-storey, 18x32 feet, disassembled and
parts marked, rafters and flooring complete $6000/
"Firm"

Logging

A constant task on most farms was "timbering", cutting trees
down, and "logging", getting the felled trunks to the proper
place. Once trees were dropped, heavy-duty wagons could
carry some of the smaller trunks, or they could be dragged out
in winter, using the assistance of frozen ground and snow.
Different types of low-slung sleds or "skids" helped, as did
plenty of power in the form of oxen or draft horses.

Logging chains of many lengths and link-weights were
necessary, and some were so heavy a 20-foot length could
hardly be lifted with one hand. The first chains were hand-
forged, with somewhat rectangular links, while factory-
produced chains had oval links. Many chains had a large
circular link at one end, for looping, and a heavy hook at the
other.

A universal tool was the log-roller or canthook, simply a long
iron hook secured near the base of a hardwood pole. Using
leverage, this could be used to move logs sideways for some
distance. Allied to it, and used when the logs were in streams or
lakes, was the log-pike, in this case a pole with a round iron tip.
It pushed logs into the desired position, and some forms also
combined a sharp hook for pulling.

An ideal compromise turned out to be the peavy, invented by
a man of that name, ca. 1870. It had both the tip and hook at
one end, so was multipurpose. Large numbers were made, and
most farms had one.

It was once common to skid logs to running water for a fast
downstream trip with the high waters of spring. For loose logs,
perhaps intermixed with those of neighbors, the wood had to be
marked as to ownership.

Log stamps or brands were used for the purpose, and they are sometimes called marking hammers. These were heavy sledge-like tools, with the owner's initials or special mark extending from the flat face of the head. Other types were much smaller, all metal, and were struck with another hammer.

If logs were kept together in "rafts", dogging chains—like log chains, but with pound-in tips—in varying lengths were employed. So also was the almost unknown log-auger, which bored holes in logs for peg insertions, so they could be fastened together.

Now that most of the great virgin forests are gone, many of the tools are no longer used or made. But all were an important part of the early tool inventory.

Values

Skid-sled, for team pulling, 7 ft. long, damaged $25-35

Log-chain, hand-forged links, end missing, 16 ft. long. . $32-48

Log-chain, 20 ft. long, half-inch thick links. $35-40

Hammer, log-marking, good handle, impressed initials
    in metal face $50-60

Log-roller, canthook with 7 ft. handle $15-20

Log-pike, 7½ ft. wood handle, iron tip $20-25

Lumberstick, graduated rod for measuring length and
    diameter of cut timber, brass tips, "F.A. Hazelton/
    Maker" $40-50

Peavy, 7 ft. handle, factory tip and hook $20-25

Log brand, handheld, iron, 10 in. long, hammer-
    struck. $20-25

Dogging chain, forged links, 13 ft. long, iron tips $30-52

Lunch Carriers

Around noon, when field workers were some distance from the farmhouse, there was no more welcome sight than the cluster of lunch carriers set in the shade. Also known as dinner pails or lunch buckets, all were handled containers designed for food storage and protection. This last purpose was very real, as farm dogs could otherwise quickly leave their owner both furious and famished.

Early wooden examples mainly carried solid foods, but later types usually had provision also for liquids. Baskets could be used, often hung from a tree branch, and empty lard or syrup cans were not only the original buckets, but cheap food carriers with tight-fitting lids.

By the late 1800's, fancy "lunch satchels" were available. One such was offered in the Peck & Snyder catalog of 1886, and it was for ". . . such persons whose business requires their absence from home during the dinner hour". In addition to two large compartments, the tea or coffee reservoir had an underneath burner for heating the drink.

Later lunch carriers resembled miniature suitcases, though by 1925 the familiar metal form had developed, being japanned with rectangular base and rounded top for a vacuum bottle for liquids.

Values

Wooden lunch-bucket, flat top, oval, bentwood sides, wooden handle $45-60

Lunch bucket, early, hand-done tin, cup on top, handle not present and no sign one was ever attached $30-38

Lunch bucket, typical shape, leather handle, light rust . $7-9

Lunch bucket, once used to ship tobacco, "Tiger" brand $26-32

Lunch bucket, round, reinforced bail, formerly held lard $7-9

Maiden Yokes

One of the more curious antiques from very early days is the unique "maiden yoke". This is a handcarved device for carrying two pails of spring water, milk or maple sap. This could be done with a certain ease, as the wood piece was designed to place full weight on the shoulders. There was a hollowed section for over that part, plus a cut-out for the back of the neck.

The user could thus stand easily upright, leaving both hands free to maneuver pails and contents. At each end of the yoke—the whole was usually 3 to 4 feet long—was a knob or groove for cords or chains to carry the containers. These in turn usually terminated with a hook fastened to the cord. The yokes were used by adults and youngsters of both sexes.

Values

Yoke, forged-iron end hooks for chains, 3 ft. 3 in. long . . $45-60

Yoke, grooved ends, all wood smooth, good condition,
    used in "sugar bush" of New England $55-70

Yoke, chains and hooks, red paint, 38 in. long $35-45

Yoke, old blue paint, hemp rope fragments, 37½ in.
    long $45-55

Yoke, wood, turned ends with wrought chains $40-60

Mallets

Mallets are characterized by large, heavy heads and a short wooden handle. The majority are made of wood, but late examples can be found with heads of reinforced hard rubber and solid cast brass. Most collectors look for high-grade all-wood mallets. The word itself probably comes from "small maul" or "maulette", which it is.

Mallets were used for tasks like striking chisels and wedges. Special mallets were used for work like knocking out the plugged hole in a barrel, and these were called "bung-starters". A great advantage of the mallet is the large head, which allowed the user to concentrate on the work and still hit the base of the tool with ease.

Heads of wooden mallets can be cylindrical, round, or square or rectangular. Some types were mass-produced using sections of heavy wood wagon wheel rims, sometimes called a "wheelwright's mallet". Very dense wood was used whenever possible, as this provided momentum to the blows and somewhat protected the head itself. Burl-wood heads were treasured, though they are somewhat scarce due to the difficulty of finding, and working, burl.

Mallet handles are of hardwood, and oak and hickory handles are common, plus some of turned wood, sometimes walnut. In cross-section handles can be oblong, rectangular or round, and many were hand-carved. Most mallets were used with one hand, and strikes were more like solid taps than very heavy blows. Collectors seek mallets with good shape, heavy well-grained wood, in unmarred condition.

Values

Mallet, carpenter's, head 10 in. high	$20-25
Mallet, head enclosed with brass tube, wood handle, 14 in. long	$25-30
Mallet, general form, good condition, 13½ in. long....	$18-23
Mallet, cast brass head, somewhat battered	$12-16
Mallet, lignum vitae wood with patina, 14 in. long	$40-48
Mallet, cylindrical head, 6 in. diameter..............	$14-18
Mallet, turned handle, round burl head, 11 in. handle .	$30-36

Maple Syrup And Sugar

Almost all early farms had enough trees—around the buildings, in the wood lot or forest proper—to include the admired maple. Not only was it an attractive shade tree and brilliant in the fall, but it provided superb wood and had a sweet sap. While other trees had late-winter runs of "sugar water", maple was always the best.

A very wide range of items associated with maple syrup and sugar-rendering was made, and all are highly collectible today. It begins in the stand of forest containing the most hard-maple trees, the "sugar bush".

Trees were drilled into or "tapped" using an auger or drill with a bit from ½ to ¾ inch in diameter. Into this shallow hole was driven a hollow tube called a "spile". These bled out drops of sap which dripped into a trough below, and sometimes also supported a sap-bucket.

Spiles were struck in with a hammer or mallet, and the first were hand-carved from elder or sumac woods. Mass-produced metal sap-directors were later available, some with hooks for holding the sap containers.

Sap collectors were originally of treen, then wood-staved, then tin, sometimes with a cover to keep out insects and debris. Such individual containers were approached by horse-drawn skids and the contents poured into a large collecting or holding tank.

This in turn was pulled to the sugar camp or a special shed in the woods known as the "sugar shack". A camp might have suspended iron or copper kettles, but the shed had large, permanently installed evaporators.

Evaporators had a system of boil-down trays, wide and flat, which gradually concentrated the sugar water into a thick, golden syrup. The best such wares were of sheet copper, set over fireboxes that maintained a fierce water-reducing heat. Galvanized-metal evaporators were also used, and some smaller tinned evaporators, round or rectangular, all now quite scarce.

In the final syrup form, in the last chamber, the syrup was carefully ladled into storage containers, these first of wood and later of metal and glass.

Throughout, a number of hand-buckets and ladles were used to transfer liquids from one vat to another. Made of wood, tin, cast-iron, copper or brass, the dippers varied in size and purpose. Much has been said about how many gallons of sugar water were required to make a gallon of maple syrup, and figures in the low 40's are usually quoted. But this varied, depending on both sap quality and desired thickness of the syrup.

Syrup became sugar through additional boil-down to the point of crystallization. Every farm maple sugar maker had a secret method for determining when the time was exactly right. Thereupon, the brown mass was poured into moulds to become maple sugar loaves and candies.

The hard mass was broken, when needed, with small clubs, or sharp-edged tongs called "sugar nippers", or a pointed, cork-screw instrument, the "sugar devil".

Values

Maple sugar stirrer, wood, holed to hang, 16 in. long... $11-14

Maple sugar strainer, tin, overall length 11½ inches ... $16-20

Maple sugar bucket, tin, japanned, ca. 1890 .... $20-25

Maple sugar thermometer, glass, copper-backed, suspension ring, floating type, 10 in. high $40-50

Maple syrup pitcher, tin, covered, wire loop finial, 6¼ in. high.............................. $42.5-50

Maple syrup pot, tin, black, spout filter, 4¾ in. high... $75-95

Maple sugar mould, carved wood, all alphabet letters, 4⅝x22¼ inches $115-135

Maple sugar bucket, orig. label, "Pure Maple Sugar",
6 in. diameter $38-45

Maple sugar form, "Vermont Machinery Co.", 200 small
tin moulds $100-140

Maple sugar mould, ceramic, yellow-ware, animal
shape, 1⅞ in. high $50-70

Maple sugar mould, tin, removable separators,
8½x15 inches $42-47

Maple sugar mould, tin, 3½x5¼ inches $5-8

Maple sugar mould, tin, 5x7½ inches $8-12

Spile, hand-carved wood, 6 in. long $1.5-2

Spile, machine turned and bored, 5½ in. long $1-2

Spile, tinned iron, machine-made, bucket hook, 4¾ in.
long $2-4

Sap trough, rectangular, wood, scorped interior,
6x13x15 inches $85-100

Sap bucket, lift-off round lid, staved, wider bottom,
16 in. high $60-80

Sap bucket, tin, attached lid, bail handle, 14 in. high . . $30-40

Evaporator, japanned tin, square, 29 in. wide $50-65

Sap stirrer, hand-carved wood, 42 in. long $16-22

Sap ladle, cast iron, large cup, 34 in. handle $30-40

Sap dipper, copper cup, wrought-iron handle, 15 in.
long $80-100

Sap ladle, tin, smith's product, 2-qt. capacity, 19 in.
long $40-55

Maple "butter" paddle, wood, some curl, 15 in. long . . . $75-90

March Nineteenth

The date probably means little to most people, but it has significance for this book. It marks National Agriculture Day, to recognize and honor American farmers, ranchers and foodgrowers. In 1981, the U.S. Congress agreed upon a joint resolution to make it an official national observance.

As the farm items collected today attest, the once-held image of farmer as rural innocent, country bumpkin or hayseed has long disappeared, to be replaced by the image of farmer as successful businessman and prime feeder of this nation and the rest of a hungry world. The astute farm-item collector might consider at least one piece marked March 19th, or literature on the subject, to help round out any grouping.

Measuring Devices

Measures are things rarely thought about but constantly used. And, judging from the way we Americans are resisting metric conversion and holding to old-fashioned standards, measures are, and were, extremely important.

We tend to think of old measures in terms of volume, wet or dry. Some of the earliest were wood-stave containers used to determine amounts of farm-grown produce, the grain measures. In early times they were crucial to the economy, for in a barter system produce was not just worth money, it was money.

Made by a cooper, the farm measures usually held from a half a peck to one bushel. The majority were made of softwood about ½ inch thick, and had a solid bottom. Sides were bound with wood hoops or heavy wires and the open top was usually greater in diameter than the bottom.

Side slats generally were joined tongue-and-groove, which provided a tight fit. This prevented even milled grain from being lost.

The simple form came about because a handle was a hindrance in a small bin. Further, the container could easily be grasped at any point along the rim. This in turn was often gently beveled if the measure was meant to be dipped and filled from a large grain hopper. Other types had squared rim edges and could have been filled from smaller containers.

Such measures were common on early farms and in two other specialized facilities. The first was the flour mill, the second the later "feed mill" which ground out various mixtures for feeding farm animals. The latter were common into the 1950's and a number exist today. Any more, most measures are metal.

Dry measures further include handwoven baskets, again in standard sizes, such as New England berry and apple baskets. Generally, a measure that was very sturdily made was used daily by a tradesperson, while baskets of flimsy material and thin wire handles meant occasional seasonal use. The latter survived only a year or two, as on potato farms.

A number of bentwood measures were also made, especially in the smaller sizes. These too had solid bottoms and were made by wrapping thin strips of pliable wood around a form. The free side-end was then tacked or otherwise secured in position.

Very similar examples exist with lids, and some are referred to as storage or pantry boxes. Many of these types were available in graduated sizes for "nesting", and were thus made more compact for both storage and shipping.

L. H. Mace & Co. of New York City offered wooden measures in their 1883 wholesale catalog. These factory-made examples were of the bentwood type, round, with straight sides, available in nested selections. Two varieties were offered, "Plain", and "Iron-bound", these with reinforcing metal strips around top and bottom.

Actual L. H. Mace sizes were from one to sixteen quarts (including two, four and eight). A nested set in plain wood, five pieces, sold for $6.50—per dozen complete sets.

Early liquid measures (some made by the "wet" cooper) were wares that were watertight, and they were as important as dry measures in rural America. Containers, from field-flask or cask to the well-bucket, all were based on the gallon and fractions or multiples thereof. Treen (tree-ware) was the earliest, and the measures were no more than hollowed sections of solid wood. Stave-constructed measures followed.

Interestingly, one of the long-obsolete liquid measures was the "gill", equal to 4 fluid ounces or one-fourth pint. Part of the U.S. Customary System for many years, the gill was often used for doling out strong spirits in a drinking establishment.

Even in the 1700's, some measures were extra well-made, of copper and brass. Pewter types, often from abroad, were also to be found. Later, inexpensive tin (tin-plated sheet iron) measures meant that any household could afford whatever type and sizes were needed. The importance of measures in rural America, even into the 1900's, can hardly be over-emphasized.

Harvested crops had to be measured, plus produce from gardens, orchards and fields. Liquid measures handled marketable water-based items like maple syrup and cider. Typical commercial units (and measures) were from pint to gallon to barrel. This last varied from 31 to 42 gallons, established by both law and customary usage.

But it was milk, and milk by-products, which required many of the measures. Often these began roughly with the fresh-milk bucket, and these were a convenient size for holding, of about 2½ gallon capacity.

Milk provided a major cash flow on the farm, whether the liquid was sold to neighbors or driven to the early dairy. Various measures established the quantity of cream and its two valuable derivatives, butter and cheese.

Cream involved liquid measures, while cheese was sold by weight, and butter by either weight or volume. Butter can still be obtained by the "stick" or one-fourth pound weight, and once could be had in container sizes up to the butter-tub.

Other rural measures include various hand-scoops, mainly used for distributing milled or whole grains, and mixing them, for farm animal feed. Hand-scoops were used in both barn and granary, where non-corn crops were dried and stored. This was either a separate in-barn or in-shed room, but more often was a complete building standing alone. Most granaries had from three to eight large bins for storage of different types and ages of field grains.

The granary would have the standard grain scoop-shovel which held about a peck, and four heaping shovelsful made a generous bushel. This in turn was often poured into a burlap bag or cotton sack for transport elsewhere.

One of the most interesting measures to be found on a few farms doesn't seem to have a name, but can be called the "chute-measure". Like most innovative devices, it was beautifully simple. It's use, however, was possible only when grain was stored in a second storey or similar upper area such as the main upper floor of a bank barn.

The floor under a particular grain bin slanted to a square or rectangular hole, which in turn had a connecting chute of some length to the floor below. The gravity-fed grain was held back by two sliding wooden slats at the lower opening, several feet apart. The volume thus contained, on three examples the writer has seen and operated, was about half a bushel.

The top slat was normally kept pushed in, or closed, the bottom either open or closed. With the bottom slat in, the top could be pulled open and the area between filled rapidly with grain. The top slat was then pushed shut, thus isolating the fallen grain. With a container placed under the chute mouth, the bottom slat was withdrawn and the grain was precisely delivered into the container, untouched by the operator. These efficient in-building measures can still be found in older Midwestern barns.

Values

Grain, ½ peck, treen, round, irregular top $100-125

Grain, staved, wirebound, one peck, 13 in. high $40-55

Grain, staved, ¼ peck, wirebound, 7 in. high $30-40

Grain, staved, double; ½ peck and one peck, 16 in.
    high $60-75

Grain, staved, two pecks, 15 in. high $45-55

Orchard, factory-splint, half-bushel, wood handles $55-70

Bentwood, one quart, solid bottom, 4 in. high $30-35

Bentwood, four quarts, solid bottom, 9 in. high $55-65

Liquid, two gills, English(?), heavy brass, handled $80-100

Liquid, ½ pint, pouring spout, copper $75-85

Liquid, one pint, pewter, handled, with spout,
    marked $90-115

Liquid, wood, about ½ quart, handled, 7 in. high $50-60

Scoop, shovel, ¼ bushel, wood "D"-shaped handle $40-55

Tin, one quart, broad spout, handled, side measure-
    marks. $10-15

Tin, quart, tunnel spout, bale handle, side marks. $8-12

Milk-Related
(See also Milking Stools)

Milk was of crucial importance on farms for the raw form was commercially valuable as a cash source, as were its three major by-products, butter, cheese and sweet cream.

Dairy or milk-related items involve three main farm areas, the milking parlor, the milk-house (earlier, the spring-house) and perhaps the summer- or out-kitchen. Most milkers had one or two favorite pails for hand-milking, and these averaged 2½ gallons in capacity. Wood, stave-constructed pails were used first, while metal pails became popular later. Some types had an extra, flat handle on the outside near the bottom, for better control while pouring.

Other dairy pails, mostly in 10, 12 and 14 quart sizes, were used to transfer milk. Even if the milking operation was not large, as on a non-dairy farm, a number of steps were involved. The milk had to be strained, cooled, stored and transported.

Folding rules or measuring sticks, partially opened. All are brass-bound with brass folding pins and hinges. Back two examples are 2 ft. long unfolded, foreground specimen, 1 ft. long, unfolded.

2-ft. rules, each $12-16
1-ft. rule $10-14

Right, Folding rule or "measuring stick" advertising; "Standard Live Stock Insurance Company / Indianapolis, Ind.".
           $8-11

Left, Sieve, common in milkhouses, with fine-mesh copper screen, removable; top, about 12 in. diameter. $7-9

Double-measure, staved and wire-bound, with half-peck compartment facing up, 15 in. high. $45-55

Measures, staved, larger with traces of old paint, wire-wrapped, soft-wood. Background, half-bushel size.
           $35-45

Foreground, peck size, plain un-treated wood. $28-35

151

Sap-taps or "spiles" for maple-sap flow; longest 9 in. long, handcarved from sumac wood. Each . . . $2-3

Well-made mallet with owner's initials on side of head; nicely turned handle, piece about 15 in. long. Head is of turned hardwood. $15-19

Milk scales, upper center and right, faced and tubular respectively, for weighing containers of milk.

Photo courtesy Museum of Appalachia, Norris, Tennessee

Some lightweight milk strainers had a very fine copper mesh for catching impurities, while earlier farmers might have strained milk through several layers of clean cheesecloth. By the early 1900's metal hand strainers, for setting above a milk can, were widely popular. Some had removable bottoms and disposable paper filters.

Milk cans, for transporting milk from farm to town or dairy, were the familiar round types with small cap and two handles on the sloping shoulders. Typical contents sizes were 5, 8, 10 and 15 gallons. Some farmers had copper letters on the sides for permanent identification, and dairies sometimes had small brass plates for their own cans.

Before the days of electrical refrigeration, the milk was cooled by ice or cold running water to stabilize it for storage. Often it was also aerated to improve taste. Milk could be air-cooled during cooler months, and milk pans of many sizes and materials were used. Typical examples held from one to 6 quarts.

Milk in the large storage and carrying cans was stirred, bottom to top, by a long rod with a holed and cupped disc at the end. Milk testing equipment (butterfat content, etc.) was available, and most operated by crank-spinning samples at high speed for separation. Some farms sold milk to neighbors, and a number of milk crocks and bottles are collectible.

Values

Milk pail, staved cedarwood, wire bail, 15 in. high $25-30

Milk pail, 3 gal. size, galvanized metal, side handle $8-10

Milk pail, tin, early, about 8 quart size, wire handle . . . $30-40

Strainer, 12 in. diameter, copper bottom screening $15-20

Strainer, two-part, used with milk-pad paper filters . . . $10-14

Milk can, 5 gal. size, with lid $12-15

Milk can, 10 gal. size, with chained lid $15-18

Milk cooler, double-wall container, two brass spigots, 19 in. high $35-40

Milk cooler pan, tin, 14 in. diameter $12-15

Milk cooler pan, stoneware, blue line at top outside $21-25

Milk cooler pan, enamelware, grey, fine condition. $18-22

Milk skimmer, perforated tin, loop for hanging, ca.
    1800's... $10-13

Milk container, "12 Qt. Liquid", cap and wire handle.. $22-26

Stirring rod, light-colored metal, 2½ ft. long $4-6

Stirring rod, solid brass, just over 2 ft. long $38-44

Milk bowl, redware, pouring spout, rim with brown
    sponging, 10¼ in. diameter $115-135

Milk bowl, white pottery, tan glaze, rim chips, 17 in.
    diameter $35-40

Milk bowl, stoneware, pouring spout, design in crushed
    cobalt, 12½ in. diameter...................... $175-200

Milk bowl, yelloware, "Sharples Warranted", 16 in.
    diameter $22.5-25

Milk can, tin, lidded, ear-shape handle, 5½ in. high ... $17.5-25

Milk can, tin, lidded, ear-shape handle, 8¼ in. high ... $20-28

Milk bowl, stoneware, 1½ gal., "Sipe & Sons/W'msport,
    Pa.", cobalt flower, 11½ in. diameter $275-300

Milk can, tin, embossed brass label, "Clark Can Co.
    Mfgrs./5 gallon/Philadelphia", 24 in. high $17.5-23

Milking Stools

The early farm family, from pioneer days on, often was limited in the number of animals they could afford. The first was likely to be a horse, for work and transportation, but the second was almost always a cow. This provided milk for drinking, for butter and cheese. If the family lived in a settlement and had small children, the cow would likely be acquired first.

Milking was done by hand, now almost a lost art. A novice was likely to be rewarded with annoyed switches of the tail, followed by powerful kicks. The milk-boy or milk-maid settled on a low stool and squirted the creamy streams into a bucket held between the knees.

Stools were kept at the right height for the user, and most, given the simple purpose, are quite plain. Round, oblong, rectangular, even triangular, most have a rough-plank top and three legs to form a tripod. Four-legged examples can be found. Most examples had holes drilled partially into the thick top or seat, and plain whittled legs were driven in.

Overall, this created a base for a temporary perch, and finishing touches were rarely included. Milking or milk-stools never achieved the status of household foot-rests, nor that degree of completion. Instead, they were kicked about by people and animals, but the stools have survived it all.

Given their somewhat plaintive origin, and the fact that these foot-high curiosities survive in some numbers, it is puzzling that they command relatively high prices. Most are considered primitives, and very collectible.

Values

Stool, block top, 2 in. thick, three whittled peg-legs....	$25-30
Stool, plank top, legs made from pitchfork handle	$30-35
Stool, plank top, four rough-turned legs, 11 in. high ...	$35-40
Stool, primitive, made from trunk and tree branches ...	$50-60
Stool, milking, oak, primitive, three legs, 14 in. high ...	$27.5-33

Mortars & Pestles

Wherever a farm might be located, the mortar-pestle combination would predate it by thousands of years. Fairly well-designed stone mortars had been used there by prehistoric Indians. And in pioneer and colonial days, mortars and pestles or pounders were a major way to produce meal, grains and powders.

One early device was the hollowed-out stump or "sump" mortar, an upright log with wooden pestle that might be fixed to a springy sapling to aid in the uplift. In line with a common use—pounding dried corn for a basic food—some such pestles were very large and the mortar itself was called a "hominy block". They were used to make raw hominy grits.

Smaller hand-powered mortars of burl or imported hardwood, usually with matching pestles, were used for many farm purposes. Ceramic and cast-iron examples abound, and were intended for grinding everything from herbs to animal medicines to charcoal for gunpowder.

The farm mortar was essentially an instrument used to pulverize some material, and they were in daily use.

Values

Mortar, hollowed from log, 32 in. high, 15 in.
    diameter $75-85
Mortar and pestle, maple, 9 in. high, 4½ in. diameter . $40-50
Mortar and pestle, cast-iron, 12½ in. high $35-45
Mortar only, white ceramic, 15 in. wide $20-28

Noisemakers

Noisemakers had several names—"alarms", "rattles", even "clackers"—but all refer to the one device. These are the early boxed wooden springs or leaves, frame-enclosed, designed to be twirled by the short straight handle that also was combined with a cogged wheel.

As the springs were lifted and released, they struck the hardwood cog with a loud snap, and the broad leaves resonated real volume. Noisemakers were once widely used and were carried by Civil War sentries and volunteer firemen. On the farm, they were used to summon neighbors, frighten burglars, drive off night-raiding predators, and celebrate weddings.

Values

Noisemaker, two-slated, painted, one slat broken...... $8-12
Noisemaker, one-slat, 10 in. high, select good woods ... $17-23
Noisemaker, two-slatted, oak frame, hickory leaves,
    16 in. high.................................. $38-46

Nutting

It's difficult to appreciate the quantity of natural foods gathered on our first farms, and the practice continues today. In big ways, farmers fed the world; in other ways, themselves. Gathering wild nuts is a good example of the latter.

In establishing a farmstead, nut-bearing trees were usually left untouched as fields were cleared, leaving selected trees to stand in stately splendor. Among such trees are the black walnuts and the so-called white walnut, the butternut. This scarce tree has an elongated seedpod, with a rich, fatty meat.

Likely to be found on farms in the South were pecans, in the Midwest, the hickory. There were varieties of these, and most produced an edible hardshell nut that sparked the palate in cold months, a gift from the year before. Hazelnuts (or filberts) were taken from low, bush-like trees.

Chestnuts were favorites for heating near coals, but were easily charred. So, iron chestnut roasters developed for fireplaces and stoves. Some examples could also be used to roast coffee.

Nuts were valued because they were nearly free, obtainable with a little effort, and "kept" for a long time. The best nuts were known to need seasoning, drying, before the real taste came out. Well-stored nuts would be good for several years.

Once the real mark of a country boy was to come to school with yellow-brown hands, the sure sign of having hulled bushels of black walnuts. Others used gloves, or pounded the nuts through holes in nutting boards, or even drove over the fallen nuts in wheeled vehicles.

Many kinds of nutcrackers can be found, from primitive to quite complex. Lever- and screw-action types are common, and a number of handcarved wood nutcrackers are collected.

Values

Carved walnut seed, folk art, into miniature basket $1-2

Chestnut roaster, iron cylinder in frame, side handle . . . $55-65

Nut anvil, cast iron, indented on top for nut, curved
    base, 5 in. wide. $25-30

Noise-maker or "rattletrap", 20 in. long, platform type
with three leaves. $50-60

Nut anvil, cast iron, knee, "Nutcracker/12-25-28" $27.5-35

Nut bag, formerly 5 lb. cotton sugar bag, marked $2-3

Nutcracker, folk art, carved wood squirrel, cracked
  body, 7 in. long................................. $17.5-25

Nutcracker, wood, man's head at end, 8 in. long $22.5-30

Nutcracker, cast iron dog, worn old paint, 11 in. long .. $20-25

Nutcracker, hardwood, with double jaws and arms,
  old $18-25

Nut-board, wood, with holes for hulling walnuts with
  mallet, 1x3 ft. $12-17.5

Nut stick, hickory, throwing club to knock down distant
  nuts, 21 in. long $3-5

Oilcans

Few more mundane items existed on farms than the ever-present oil can. They range in size from small to very large, and placed lubricating drops on knife edges, lawnmower wheels, horse-drawn machinery, in fact, almost any moving metal parts. It prevented both rust and excessive wear.

Oil was obtained in quart and gallon sizes, and was transferred to the oil cans. Most collectible varieties have a large, flat base, conical or cylindrical bodies and spouts from 1 inch to 1 foot in length. To reach difficult places, the spouts might be straight or angled.

Other than bottom-pressure types, which put out oil by thumb movements, others had small pumps. Two types were the variety with thumb lever atop the handle and a type with palm-squeeze lever aligned with the handle.

Collectors prize copper oilcans and those with brass fittings, such as valve parts or spout-tip. Too, examples with maker's marks are of interest.

Values

Can, coated tin, 3 in. base, 4 in. spout $1.5-2

Can, maker-marked on side, "Dietz", 9 in. high $3-5

Can, tin, thumb delivery system 13 in. high $5-7

Can, copper, 7 in. high, unmarked $12-15

Can, tin and brass, handle-squeeze oil delivery........ $15-20

Can, brass and copper, polished, 8 in. high $25-35

Above, Orchard item, sprayer container, made of rolled sheet copper throughout, 13 in. high. $28-35

Above, double cherry seeder, central hand-cranked ridged disc; "Cherry Stoner / No. 117". $28-35

Cherry-seeder, ridged wheel and hopper type, clamp base, "Bogan and Strobridge". $18-25

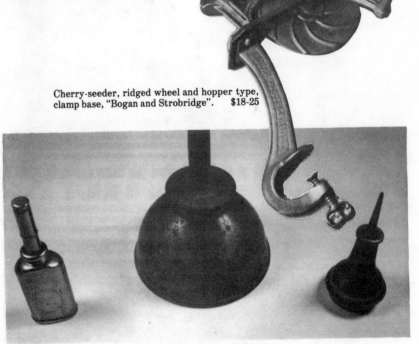

Oilcans. Left, chromed with protective cap, 3 in. high. $3-4
Center, canted long spout, copper-plated, "Eagle". $5-6
Right, spun and sheet brass. $6-8

159

Log-constructed animal pens and enclosures. Lower left center, the ramp used to load small animals for transportation.

Photo courtesy Museum of Appalachia, Norris, Tennessee

Above, A common farm advertisement, signs now disappearing.

Ox-yoke key, hand-wrought iron, 6 in. high.
$7-10
Photo courtesy Fairfield Antiques, Lancaster, Ohio

Ox-yoke with twin ox-bows, bentwood, one sprung from hole.
$35-45

Abandoned barn and wagonshed on Midwestern farm.

Hogshed, early, with wood-shingle roof.

Barn, with projecting second storey which serves as protection to either stock or machinery.

Photo courtesy Museum of Appalachia, Norris, Tennessee

Subterranean structure with different regional names. In the Appalachians, it is the "dairy" for milk and its by-products; in the upper Midwest, it is the root cellar; further East it might be called the cave cellar. Always, it was for cooling and storing various foods.

Photo courtesy Museum of Appalachia, Norris, Tennessee

Wood supply, stacked in lean-to type of woodshed.

Photo courtesy Museum of Appalachia, Norris, Tennessee

Orchard items
(See also Apple-Related)

Husbandry—taking care of the land—certainly included the selected trees grown on the farm for fruits. Always, the apples were dominant, and fully half the two or three dozen trees were usually apple varieties. Aside from the named types, apples were classed as those that tasted best fresh, or cooked well, or stored without spoiling.

Additional trees would include plums, cherries (sweet and sour), pears and peaches, plus anything else the owner had a taste for or thought would produce well. A number of antiques and collectibles are related directly or indirectly to the early orchard.

Grafting knives are of two main styles, and the more common has a large curved blade. Many farm pocket knives with folding blades had a sturdy hook-like blade for cutting and connecting treestock. Tree care involved a range of clippers, both hand and pole, for pruning limbs. For larger branches, hand and pole-mounted saws were needed.

Sprayers include pump-types, canisters opeated by compressed air, and larger hand-pulled and wheeled examples. Fruit-pickers are surprisingly abundant, and were used to harvest fruits beyond reach from ground or ladder. Most had metal plucking "fingers", and a small basket for from one to half a dozen fruits. At least one innovative picker had a long canvas tube that went from the blade or fingers to a bag carried by the worker.

Ladders definitely used in the orchard are not common; they tend to be wood, tall, "A"-shape in profile, and have an extra-wide base on the step side. the support side sometimes terminated in a single point, giving tripod stability.

Very many orchard containers exist, both the cotton or canvas shoulder-slung bags and the gathering or orchard baskets. Most of the latter, whether handmade wood-splint or a factory product, have a peck to a bushel capacity, two handles, and durable construction.

Once the fruit had been picked and gathered, it could be further processed. Cherries and peaches did not store well, and had to be canned or made into preserves quickly. Apples, however, could be placed in the fruit or root cellar, partially below ground, and insulated against the cold.

Cherry pitters or stoners are designed to remove the seed from the fruit, and primitive and patent devices both are common. Most were made to be mounted on the edge of a plank or table, and all are both interesting and efficient. Wheel and plunger models are the most common.

Values

Grafting knife, wood handle, 10¼ in. long $10-13

Grafting blade, contained in folding pocketknife, wood
    handle, open length, 6¼ inches $12-16

Pruning clipper, factory-made, 14 in. long $10-15

Pruning saw, curved blade, 17 in. long, mounted on 5 ft.
    pole $15-20

Pruning hook on pole, lever-operated at base $17-22

Pruning saw, burl handle, 22½ in. long $35-45

Sprayer, early, carry-tank with brass fittings, copper . . $60-75

Fruit picker, head only, iron and wire, metal fingers,
    14½ in. long $11-14

Ladder, "A"-frame, steps on one side, wood, 10 ft. high $35-45

Fruit press, cast iron, wheel and thread plunger, 14 in.
    high $35-45

Fruit press, wood, fulcrum/lever plunger, staved con-
    tainer, old, 22 in. high $90-115

Cherry pitter, cast iron, table-clamp, "Rollman Mfg.
    Co.", 11 in. high $22.5-30

Cherry pitter, twin-pronged plungers, tabletop model,
    8 in. high $28-35

Cherry pitter, corrugated disc type, crank handle, 10 in.
    high $26-32

Sign, "Farm Produce/Farm Prices", 13½x38¾ inches . . $80-95

Fruit-gathering basket, two-handled, 9½ in. diameter . $59.5-70

Oxen

Even before horse and mule power occurred on the farm, pairs of oxen did much of the heavy work. They were in fact preferred for certain jobs like plowing, logging and ice-harvesting. Despite popular belief that oxen were a species separate from cattle, an ox is simply a large, neutered bull.

A surprising number of ox-related items can be found by the diligent collector, and such pieces carry a certain ancient aura. This may be because while oxen are common today in beef herds, they are no longer regarded as work animals.

Values

Ox-yoke, heavy bent wood, old red paint, weathered .. $80-95

Ox-muzzle, of plaited wood strips, used to prevent oxen
    from eating field grain while working, 10 in. high . $40-50

Hoof-trimming blade or "buttress" for oxen, curved
    blade with end and side handles, ca. 1835, 15 in.
    long $35-45

Ox-goad, 6-ft. stick with sharp tip, whittled hickory ... $10-15

Ox-lead, heavy cast iron nose-ring, for guiding work ox,
    3½ in. diameter $12-15

Horn-caps, pair; wood, holed balls used to protect people
    and animals from being gored, late 1700's $24-30

Ox-shoe, two part, wrought iron, pitted, 5 in. long $5-8

Pest Traps

Build a better mousetrap—and someone comes up with one
even more efficient. Rats and mice caused incalculable losses on
yesterdays farms, and remain a problem to this day. They ate
quantities of corn and grain, fouled areas, and gnawed holes in
everything from work gloves to concrete building foundations,
this a specialty of the Norway or common brown rat. They
killed baby chicks, spread disease, and were unsightly on the
well-kept farm.

Adverse reaction to such vermin ("varmints") culminated in
the great farm "pest hunts" of the mid-20th Century. Then,
rural youngsters, with the support of school-based
organizations, engaged in contests to bring in the most sparrow
heads or rat-tails. A bit gruesome sounding now, such massive
drives were quite successful.

All along, a common practice was to have traps in key places
in many of the outbuildings, the chicken house, granary, and
barn. The widest range of collectible pest traps were used
against mice, as these had to be custom-made. As it was, traps
used to capture smaller fur-bearers like muskrats were used for
farm rats.

Several hundred different mousetraps evolved, including one
variety that first caught mice and then channeled them to a
Ferris-like wheel where they ran endlessly, for children's
amusement.

Found everywhere were the Victor-brand "3-Way" traps, in both mouse and rat sizes. The heavy wire mechanism was stapled to a rectangular wooden base. They never wore out and they seldom missed. There is growing interest in all forms of farm pest traps, and some collectors specialize in them.

Values

Mousetrap, metal, "V"-shape round jaws, 3 in. long . . . $2-4

Mousetrap, primitive, mouse trapped and struck by
    falling hardwood block, well-made, 14 in. high . . . $50-60

Mousetrap, tin and wood, "Household/Erie, Pa.",
    5¼x6 inches $40-50

Mousetrap, wood and tin, early, intricate entrance with
    moving floor and door that closes behind mouse . . . $55-65

Mousetrap, four-way, round, black plastic wheel with
    four entrances, sprung-wire catches, ca. 1945 . . . . . $7-9

Rat-trap, galvanized wire, trap-door, "humane", live-
    catch, 10x10x18 inches $12-15

Rat-trap, primitive, large box with falling top and end,
    triggered by bait on long wood trigger, 28 in. long . $45-55

Rat-trap, converted muskrat single-spring trap by filing
    pan sear-catch to be ultra sensitive $5-8

Pigeons

Pigeons were once esteemed for food, and squab (the young bird) and pigeon-pie were considered delicacies. Large and wealthier farms might have a special building for raising pigeons, sometimes called the "dove-cote". There was usually enough loose grain around the buildings and in the fields so the pigeons did not have to be special-fed.

More likely, and much easier, was the common practice of putting nesting boxes at one end of the barn or other large out-buildings, up near the eaves. Holes were made in the siding so the birds could come and go. Such a setup was known as a pigeon-loft, and the nesting boxes, the pigeon-holes.

Values

Nesting boxes, pigeon loft, set of twelve $22-28

Pigeon carrier, tin, old green paint, 14 in. long $35-45

Pigeon carrier, woven wicker, single handle, 15 in.
    long $30-38

Planes

While not all farm boys could handle every piece of equipment in the toolshed, most quickly became proficient in using planes. These were used with adjustable blades to shave and smooth wood, and there are dozens of plane types.

Early varieties were all-wood except for the steel blade or "iron". Later came the metal-bodied planes, some still retaining a wood handle. Beech with good grain and hardness was often the preferred wood.

Most planes, no matter the size, were used with both hands, one to guide the work, the other to provide continuous force. Planes can be classed as to blade-edge configuration, each being designed for different cuts in the wood.

Common plane types include: Tongue and groove pairs (or "match"), rabbet (bull-nose and jack), smoother, jointer, skew, hollow, round, and others. A simple wooden wedge held the iron firmly in early examples, while metal planes had blades secured by screws, catches, or both.

Values

Plane, wood, 15 in. long	$20-25
Plane, with reversible rockers, metal, 10 in. long	$45-55
Plane, wood, handle at rear, 10 in. long	$24-28
Plane, adjustable molding, beautifully made from exotic wood; positioning screws with ivory knobs, 12 in. long	$127.5-150
Plane, "Auburn Tool Co.", with thistle mark, 16 in. long	$20-25
Plane, dado, "Ohio Tool Co.", brass knob, 9 in. long	$34-40
Plane, "Deforest/Birmingham", 10½ in. long	$20-25
Plane, "D. Kennedy", modified with extension, 16 in. long	$30-35
Plane, "D. R. Barton/Star Trade Mark", 9 in. long	$16-20
Plane, "Ohio Tool Co.", 16 in. long	$18-22
Plane, scrub, "Stanley #10", 1½ in. blade, 10 in. long	$25-30
Plane, scrub, "Stanley #40/1896", 10 in. long	$35-40
Planes, tongue and groove, pair, "Sandusky Tool Co.", each 13 in. long	$80-95
Plane, moulding, adjustable fence, blade gone, 9¼ in. long	$35-50

Planters
(See also Corn-Related)

Beyond seeders that scattered small grains and the several types of corn-seed planters, another hand-operated planter type was used. This was the tubular potato planter, with planting head that automatically opened when the seed potato section was at the proper soil depth. Cut potatoes were dropped down the hollow handle and were efficiently deposited in the ground.

Another hand-powered device could actually set out plants or seedlings, from strawberries to tobacco "starts". Not only were the plants set well and firmly, but the device watered and/or fertilized at the same time.

Values

Potato seeder, wedge head, solid wood handle 3 ft.
    long $20-30
Plant-setter, metal with water reservoir, "Master's" $30-40

Playthings

A wide range of make-it-yourself playthings developed on farms before store-bought and mail-order items became easily available. Slingshots were handcarved from sapling forks, the rubber often obtained from an old auto innertube. Before the popularity of airguns, bean- and pea-shooters could be made from a section of pipe or tubing.

Whistles could be carved from a piece of willow branch, and dolls made of cornhusk with dried-apple faces. Even the heavy lower portion of a cornstalk, with a strip of tough outer layer raised and peg-bridged, could be turned into a crude violin. The bow was the rib of a corn-leaf, and tone depended on the width of the raised layer.

Adults had playthings too, and a common game was "barnyard pool", or horseshoes. Before the game developed into an actual sport with rules, it consisted of lofting worn-out horseshoes at metal stakes.

Values

Slingshot crotch, hand-carved, 6 in. high $7-9

Whistle, part of bark missing, two-tones, old $6-8

Pea-shooter, wooden drilled tube, 11 in. long $5-7

Pop-gun, cane handle, dowel pusher, 9 in. long $8-11

Doll, cornhusk, apple face, deteriorated condition, 8 in.
 high $10-14

Horseshoe, worn, once painted and hung as a good-luck
 symbol $3-4

Pliers

Pliers have pivoting jaws for holding or bending, and the word means a tool used to ply a trade. Farm pliers were necessities in the shop or tool box on any field equipment, as they had a wide range of uses. Many had only one "spread" or size the jaws could be expanded to, but both the slip-joint and the pump-joint pliers allowed the jaws to be spread futher.

To increase utility, other pliers had wire-cutters or shearing blades, even screwdrivers and reamers at the handle tips. Probably no other single farm tool was used for so many different purposes.

Values

Pliers, flat-nose, maker-marked, steel, 5 in. long $4-7

Pliers, box-join, marked, toughened steel, 6 in. long . . . $5-8

Pliers, long-nose, "Fult--", 5½ in. long $3-5

Pliers, combination, "Diamond Edge", 9 in. long $11-13

Pliers, saw-set, for angling teeth, 13 in. long $12-14

Pliers, stamping, "H. J. Schmidt/Wisconsin", 10 in.
 long $15-18

Pliers, steel, Winchester-marked, 6 in. long $19-25

Plummets

Plummets or "plumb-bobs" are conical or top-shaped weights secured to a line, and were used to determine the vertical alignment of something, or depth. They were part of the tool-kit of most farms, and could be used for everything from setting a corner post for a building frame to digging a straight and narrow well. Of steel or brass, they combined weight with a small mass to achieve the best gravity-drop.

Values

Plummet, iron, cord eye at top, 3½ in. long $8-12
Plummet, brass with steel point, mercury-filled, 6 in.
    long $40-50
Plummet, steel tip, top unscrews for lead lines, 3 in.
    high $18-22

Postcard Photos

In the early 1900's a combination of rural pride and enterprise made a happy marriage, and the progency survives to this day. Some farmers with valuable holdings in land, livestock, and buildings wished to make a statement. In a low-key way, they wanted people to know who they were, where they lived.

Photography at that time had advanced far beyond the tintype and clumsy laboratory stages. Lightweight cameras, foolproof film and inexpensive image reproduction were available. Traveling photographers stepped into the economic picture and toured the countryside, concentrating on the scattered rural households. Also, some town or studio photographers did this.

The common farm photo postcard was a general overview of the house, barn and outbuildings, plus nearby fields. Often taken from some distance, family members or workers were rarely in evidence. Such photos were largely done in black and white, as professional hand-tinting (often done in Germany) and printing were expensive.

Unfortunately, few are ever marked as to place or time, though some—coming down as family records—can be identified. Most were purchased in bulk by the farmer and were run off in the standard postcard format. The back was the farmstead, the front had the typical address/message setup.

Farm postcards, many of them mailed and with writing, are still reasonably priced and can be found at farm auctions and in Turn-of-the-Century albums.

Plane, exotic hardwood handle, "Stanley / No. 45".
$60-75
Photo courtesy Tilson Collection

Rat-trap, "McGill Metal Products / Marengo, Illinois", 6 in. long, wooden base has hole for nailing to floor. $7-10

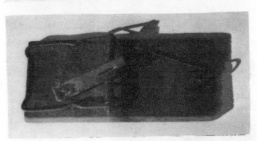

Mousetrap, wood base, "Can't Miss / McGill Metal Products Co.". $3-4

Rat-trap, wood and metal construction with revolving drum. Lower center and right, two types of wooden maul.

Photo courtesy Museum of Appalachia, Norris, Tennessee

Skew rabbet plane, about 10 in. long, probably boxwood, orig. ornate wedge and early iron. Maker-marked. $24-30

Plane, marked "New York Tool Co. / Thistle Brand / Auburn, N.Y.", 8 1/8 in. long. $28-34

Small wagon, these made for boys by older men. Though mainly toys, they were also used to haul firewood, vegetables and much else; they are typically about a yard in length.

Photo courtesy Museum of Appalachia, Norris, Tennessee

Values

Midwestern farm, distant perspective, shows out-
buildings $2-3

One-horse buggy, with two dressed-up young men $3-4.5

Rural families, many people gathered for picnic or
reunion $1.5-2

Worker, cutting with scythe $2-3

Old gentleman by out-building door $2-3

Rural family, gathered in hayfield, horse-drawn wagon
and load of hay in background $3-4

Potatoes

According to an old folksong, early American farm families mainly subsisted on "Potatoes and meat, meat and potatoes". This isn't far from the truth, for the first root crop put in was likely to be the pale, fleshy tubers. So long as the pigs didn't nose them out, the family had an assured food staple.

Shovels and forks helped with harvesting in early days, but later commercial operations employed bigger equipment. Many of the pieces—like powered vine-beaters and the tractor-pulled sprayers and potato-diggers—are too large for the average person to collect. They remain a target for the farm-equipment collector.

However, smaller pieces have both charm and appeal. One would be the horse-drawn potato-plow, with a large, comblike set of teeth angling into the soil. Another is the grading table, and early examples sloped so that rolling potatoes would pass over holes sized from small to large. Potatoes dropped into appropriate bins or baskets beneath same-sized holes, with "seconds" or "culls" removed by hand.

Other collectibles include any early bag-tags, usually of stamped or printed cardboard, and the burlap bags themselves if from defunct operations. A little-seen device is the wood-handled metal-armed hook, the tie-twister, used to wrap looped copper wires around the neck of smaller paper bags. These closed bags of the one- and two-peck sizes, and such items were used until around 1950.

173

Values

Potato-plow or digging rake, metal teeth, twin "walking"
 handles, 3 ft. handle length $55-75

Grading table, hardwood, 11½ ft. long, four different
 hole sizes $60-80

Bag-tag, printed thin cardboard, with string-hole,
 1940's. $1-2

Tie-twister, 8 in. long, canted iron with end-hook, for
 copper tie-loops, unusual. $8-11

Slicer, mechanical, for cutting seed potatoes, lever-
 operated, 2½ ft. high $55-70

Powderhorns and Flasks
(See also Shooting Accoutrements)

Few items were more common or more necessary in early
rural America than powderhorns. The indispensible companion
of any muzzle-loading rifle or shotgun, they were hung on
walls, suspended by shoulder straps, tucked away in hunting
bags.

Powderhorns were widely used from the late 1600's into the
1930's. The late dates prove that muzzle-loading rifles were
well-used into the Depression Era. This was probably because
old muzzle-loaders were both available and inexpensive to
shoot.

The average powderhorn is 6 to 10 inches in length, though
larger and smaller specimens can be found. This size was the
typical carrying horn. Two other types exist, both
comparatively scarce. One, perhaps the hardest to find, was 12
to 18 inches long, and was kept as a supply horn. Used with a
small funnel, it filled the carrying horn.

Another powderhorn was much smaller, 3 or 4 inches long,
and it is one of the earliest varieties. It was used in conjunction
with the regular horn, but only with flintlock rifles. Called a
priming horn, it was used with finely ground black powder to
fill the flashpan. Ignited by the strike of flint against the steel
pan cover, this touched off the main powder charge via a small
opening, the touch-hole.

Some sellers feel the priming horns are only small or under-
sized powderhorns, but they are distinct bits of Americana.
Priming horns were used until about 1825 when caplocks came
in.

Actually—unless the shooter did not know his business or it was an emergency situation—the carrying powderhorn was not used to pour powder directly into the rifle muzzle. A sized measure, often attached to the horn, was used. This guaranteed that the same charge was put in each time, insuring accuracy. And there was no danger of bursting the barrel.

This procedure also prevented a freak and deadly accident. Sometimes still-burning particles were left in the barrel after a shot. Then, if a fresh charge were poured in and ignited, this caused the entire powderhorn to explode in the user's hand, a few inches from the face. Accounts of such instances can be read in the occasional early journal.

Besides the horn portion, which was sometimes worked so thin the powder level could be seen against strong light, carrying powderhorns had three other features. There was a large and round or oblong base plug, permanently fastened with tacks or small nails to the larger rim of the horn. The base plug had a protruding knob or staple, to which ws fastened one end of a carrying strap.

The other end was secured to the tip of the powderhorn, often at a thin groove an inch or so from the pouring tip. The hollow tip was in turn closed by a wood or bone stopper, this small item itself sometimes carved in a pleasing manner.

In Eastern and Midwestern parts of the country, powder-horns were typically made from cattle horns. In the Western Plains region, buffalo (bison) horn was also used.

More elaborate horns exist. These may dispense with the traditional wooden base and stopper, and have those areas overlapped and reinforced with German silver or copper. In such cases, the horn itself may have been heated and moistened to force it into a rectangular (in cross-section) shape. This made the horn hang closer to the user, was more artistic-looking, and must have been something of a status symbol.

Even more sought-after are horns with incised patriotic or other scenes and slogans. Some museum-quality horns, often medium-large in size, bear maps of early settlements or explorations, complete with long-ago place names. A few rare pieces, often dating in the 1700's, were made of ivory, with the entire surface carved in low relief. Scrimshanders' work of this quality may appear in museums and at major collector auctions.

As with anything collectible, there can be problems. Often the powderhorn is a plain, but good, piece. But after about 1950 numerous reproduction muzzle-loaders were sold to "smoke pole" buffs. With the modern flintlock and caplock weapons came powderhorns, and these are now beginning to show a quarter century of age.

Powderhorns began to be replaced by powder flasks in the mid-1800's, for several good reasons. They better protected the black-powder from moisture, and were more ornate. Many types were made with embossed scenes on copper, brass or pewter, with others were metal and leather-covered. They were turned out by the thousands.

Values

Powderhorn, priming, 3¼ in. long, plain	$15-20
Powderhorn, storage, 15½ in. long, with funnel	$75-90
Powderhorn, good condition, with measure, 9 in. long .	$18-22.5
Powderhorn, notched small end, 10 in. long	$15-20
Powderhorn, wood base (turned), 11 in. long	$25-30
Powderhorn, rare, folk art map of New England, etched details, ca. 1635-1700, 15 in. long	$700-900
Powderhorn, "Death Before Dishonor", 13 in. long	$65-80
Powder flask, "Am. F. & C. Co.", large, leather covered	$40-50
Powder flask, brass, embossed hanging game scene, "Dixon"	$85-100
Powder flask, copper, "G. & J. W. Hawksley", 8 in. high	$55-65
Powder flask, copper, embossed with ivy leaf design . . .	$65-75
Powder flask, "Dixon/Sheffield", leather-covered	$35-45
Powder flask, small, shell pattern	$40-50

Pulleys

Grooved wheels held in a frame, pulleys helped change the direction of force applied to the rope it held and guided. Farm pulleys ranged from lightweight (above the farm well) to heavy, used in the hayloft. There were other uses, as in extracting small stumps, old fenceposts, or in lifting loads to upper levels of farm buildings.

Primitive cannon toy, about
7 in. long. Barrel and
"chassis" hand-carved, wired
wheel is an old spool. This is
a charming bit of Americana.
$12-15

Above
Miniature hay rake, handmade, reported to be over a century old.
Photo courtesy Museum of Appalachia, Norris, Tennessee

Powder flasks, left and center, two plain, two with embossed designs, one at center
leather-covered. Range per each . . .
$65-100-plus

Photo courtesy Garth's Auctions, Inc., Delaware, Ohio

Fine barn pulley, single-wheel of hard-wood, cast-iron open framework, pivoting closed eye, over 12 in. long. $11-14

Pulley, for ¾ in. rope, hardwood sides, wood roller, cast-iron eye, 9 in. high.
$11-15

Above, Bailey iron-bodied plane, adjustable, wood fore-knob and handle, original "iron", fine condition. The Bailey firm was acquired by Stanley. $25-30

Above
Puzzle, hand-made, six pieces of interlocking wood, 3 in. at any dimension.
$7-9
Photo courtesy Fairfield Antiques, Lancaster, Ohio

Powderhorns, top example 10 in. long.
Top, dated (18)"88" $22-27
Lower, leather thong strap $20-25

Energy applied to the rope was, early, by horse or mule, later by tractor or power-driven winch. Pulleys seen today in antiques shops are mostly hay-related, designed to take the long ¾ to one inch manila hay-carrier rope. The most valuable are the all-wood examples, followed by later wood-and-metal types.

The more common pulley is the single wheel, then the twin and triple-wheel pulleys. Each has a heavy hook or eye by which it was fastened to a barn floor, roof or beam. Many old barns have upper lofts containing a few pulleys, suspended far beyond reach of the casual collector.

Values

Pulley, wood, single wooden wheel, wood frame	$8-12
Pulley, wood, double wood wheels, refinished, with eye	$16-20
Pulley, wooden wheel, cast-iron openwork frame	$10-14
Pulley, wooden wheel, cast-sheet-iron frame	$8-11
Pulley, iron wheel, cast-iron plain frame	$7-9
Pulley, water-well, for ¼ in. rope, open-spoked	$9-12
Pulley, general-purpose, for small rope, support hook	$7-10

Punches

Though simple in design, punches had a place on every farm. Of iron or steel, they are long and tapered, with a small and large end or head. In use, the hand steadied the punch, with its tip placed against the object to be displaced. The head was then struck with a heavy hammer.

While the term "punch" also refers to objects that made holes (harness-making) or stamped a surface design (metal-working), as used here it is a device to remove bolts, pins or rivets. Such punch use was a key to repairing broken equipment of all sizes and types, especially knocking out damaged metal parts for a replacement. Most punches are of hardened steel, and are thus nearly unbreakable.

A related class of little-known punch-like tools are "eye swedges". Of metal, they were used to repair the broken handles of certain tools. Often the break was just at the head, leaving a securely wedged socket inside the metal portion. This occurred with hammers, axes, sledges, and so on. Eye swedges neatly knocked out the old handle section so a new one could be fitted.

Maker-marked punches are the most sought-after, also unusual or special-purpose types.

Values

Punch, iron, ½ in. diameter, 5 in. long $2-4

Punch, steel, 1 in. diameter, 9 in. long, tapered tip $3-5

Punch, adjustable, "Apex", 7 in. long $8-10

Punch, "Samson #4", pliers grip, 7½ in. long $13-15

Punch, early, string-wrapped at center, head "mush-
    roomed" $1.5-2.5

Eye swedges, set of six; handmade, iron, average is 4 in.
    long $100-120

Rail-Splitting

To clear the land of trees and make room for fields, early Americans used one to surround the other. They cut logs of oak, hickory and locust and split them to make long, thin sections for rail fences.

Unfortunately, a wrong impression has developed as to how logs were split into rails, with illustrations showing such figures as Abraham Lincoln using an axe. (If Abe had actually done so, he would have been the first President to lack the usual number of toes.) Axes stick, or have the pounding poll battered to distort the eye, or strike glancing blows that are dangerous.

Instead, the rail-splitter used hardwood triangles called "gluts" to start and widen the splits. These were struck with one of two instruments, either the "maul" or "beetle". The maul was simply a long, heavy club, and the name survives as a catchword for devastation. The beetle was a handled tool with a large, cylindrical (often iron-bound) head, much resembling those used by old-time circus workers to pound in stakes for the Big Top.

After the Civil War, iron or steel wedges came into wide use, struck with the sledge, which had a heavy, cast-iron head. This combination is still used today.

Values

Glut, 8 in. long, badly battered, old $2-3
Maul, 39 in. long, well-used, hardwood $5-10
Beetle, iron-bound head, newer handle $7-13
Beetle, burl-head, 30 in. long, fine condition $15-25
Iron wedge, factory made, 6 in. long $2-3
Iron wedge, early, with owner's initials on side $5-8
Sledge, old, cast head, 27 in. hickory handle $5-9

Roadside Markets

Beyond the Grange movement, Farmers Exchange and other agricultural organizations, some farms sold produce beside the road. The goods were generally neither really field-crops nor garden items, but fast-moving fruits and vegetables that passersby were likely to purchase. The source was usually the "truck patch" or commercial garden or the orchard.

Sweet corn, melons, cherries, peaches, apples, potatoes—all were offered at good quality and reasonable prices. There was something about buying directly from the land that appealed to "the traffic", and many a farm child received a first taste of the free enterprise system in this way. Such operations usually developed local reputations from extra-good eggs to grown things.

Values

Cash-box, four bill slots and round change compartment,
    one-piece wood, 13 in. long $11-15
Sign, "Fresh Farm Eggs", white on black-painted wood,
    17½x23 inches $9-11
Fruit-stand display pyramid, for farm wagon, about 9 ft.
    long, three levels . . . . $-No bid
Sign, "Field-run Melons/Sweet/10¢ Each", faded paint
    on whitewash ground, 2 ft. high $25-30

Routers

Fairly intricate, routers have special blades to dig or scrape out wood; used with a fence, this could be done a certain distance from, say, a board edge. Routers were important for working with doors and windows, also for making furniture.

Values

Router, early, iron, cuts ⅞₆ inch wide................ $60-80
Router, "Stanley #71/1884", 7 in. wide $25-30
Router, "Stanley #71", with depth gauge and fence,
   6½ in. wide $30-35
Router, "Stanley #71½/1901", 7 in. wide $20-25

Sack Holders

Filling grain sacks by hand was a two-man job, one to hold the sack or bag open, the other to scoop and fill. One man could do the work if he had one of the many kinds of sack holders, which both held the sack up and the mouth open.

Some holders were designed to be nailed to a wall, others to be hung over a slat of the grain bin or the side of a wagon. Others had their own platforms and poles, and could be set up anywhere.

Values

Sack holder, cast-iron, wall-type, two projecting curved
   arms...................................... $15-20
Sack holder, platform type, two pole-mounted spring-
   powered holders and openers, factory-made $30-40
Sack holder, farm-made, galvanized funnel-type top ... $20-25
Bag filler, pine, adjustable hopper holds grain sacks, old
   worn Amish blue paint and good honest wear, 52 in.
   high $95-125

Sale Bills

Sale bills are the printed poster-like announcements of farm sales, which appeared in country stores and almost any place of public notice. They are common today in newspapers, but larger-size versions, often on colored paper (standard style was black print on white paper or light cardboard) were once rural advertisements of a coming farm sale.

At the top, the sale bills or auction broadsides gave the particulars of who was selling what, when and where. Top billing went to farm buildings, machinery, livestock and acreage to be sold. The bulk of the bill, however, was often taken up by a general listing of farm items, and this in itself can be an examination of old, period items.

The bottom of the bill usually listed the auctioneer(s), any special terms, or, "Terms on day of sale". There might also be comments on "eats" and restrooms or "comfort stations", plus the ubiquitous sales notice, "Not responsible for accidents".

Sale bills are valued as to age and condition, and the number of farm items listed. Collectors seek own-county or state-of-residence bills, or those of well-known estates. While large numbers of such bills were printed, few were ever saved to survive today. This is an under-collected field with considerable potential.

Values

Farm-sale bill, Iowa, fold-marks, typical listings, year
    1937 . . . . $12-16
Farm-sale bill, Ohio, framed, listings heavy on household
    goods, year 1923 . . . . $16-20
Farm-sale bill, Pennsylvania, creased, faded, 1929 . . . . $10-14

Salesman's Samples

The traveling salesman has long been part of the humorous folklore of, or about, rural America, usually combining in story form rustic wit and city-slicker ways. Such ever-moving business agents might sell directly to farmers or offer examples, samples, of their wares. These could be purchased on-the-spot or ordered to be delivered later. This worked well when objects were small and light, but such was not always the case.

Correspondingly, salesman's samples developed, these being miniatures of the real object, smaller-scale representations of the manufactured product. Almost always, these duplicated in every way the actual product, complete with working parts and company name.

While it is yet easy to confuse such samples with certain play-things, miniatures, even patent models, a number of salesman's samples can be tied to the family farm.

Small or fragile sample-containers were usually carried and/or displayed in a protective case. These were often small quantities of material like chemicals or even lubricating oils. Other known examples are small-scale recreations of wooden farm gates, agricultural implements, wood-burning stoves, and—stretching the imagination—small stone burial crypts.

Collecting salesman's samples of farm equipment and implements is a whole field in itself, one with growing appeal.

Values

Salesman's sample wooden gate, on board base, workable,
    about 2½ ft. long, all parts present $100-125
Salesman's sample oil-grades case, with most labeled
    bottles and contents present $35-45
Salesman's sample corn cultivator, brand-named, with
    hand-written documentation re. actual use period . $125-150

Scales

The smallest scale in general farm use is probably the egg-scale, measuring weights from pullet (smallest) to double-yoke sizes. The largest scale, in wide use by 1875, can't really be collected by the average person because it was building-sized. These were the giant Fairbanks scales, some also put out by other makers.

The Fairbanks drive-on scale had a shed or scalehouse built over it, and the weighing surface was also the shed floor. The weight of, say, a wagon loaded with sacked grain could easily be determined. the empty wagon was driven on and the weight noted. The full wagon was later weighed, and the first total subtracted from the last. Scalehouses can still be found on some early farms, though they are now mostly used for other purposes.

Smaller platform scales were used on every farm, and could handle weights up to about 400 pounds. Portable when carried by two men, they had a square or rectangular weight-surface and a wood-encased pillar that supported a scaled bar with end-weights. Many variations existed, and they weighed everything from feed mixes to partly filled milk cans. Of all scales, these are the most farm-related. Such scales could be preset to allow for the weight of a container, and thus noted only the contents.

Spring-type scales were widely available and inexpensive, though the admonition marked on many—"Not Legal For Use In Trade"—was a point well made. They were never totally accurate, as age and even temperature made performance less than complete. Nonetheless, they exist in nearly every imaginable size, with capacities from 20 to 300 pounds. Scaled faces are often of sheet-brass, though some have sheet-iron.

A simpler, very early scale was the arm-balance type or "justice" scale. Center-pivoting, of wood, metal or both, the object to be weighed was placed on one suspended platform and weights were put on the other. Weights themselves for this scale are interesting, being first stone, then cast-iron or cast-brass.

Another common farm scale is the suspended steelyard, sometimes referred to as a "stillyard". A very old type, it dates back to Roman times. Hung off-center, it has a long scaled arm and a shorter arm to hang the object to be weighed.

A sliding weight counter-balances the object and measures the weight on the arm. These are sometimes offered in Plains states as bison-hunters' hide scales, at high prices, but they remain general-purpose scales with many uses.

Values

Balance scale, tin platforms, chain suspensions	$28-34
Beam scale, heavy duty, taking weights up to 320 pounds, without weights	$25-30
Brass-faced scale, polished, weights up to 28 pounds, "Frary Improved", 1 in. wide	$27-33
Brass-faced scale, "Frary's Imported Spring Balance/ Warranted", up to 25 pounds	$26-32
Brass-faced scale, "John Chatillon & Son/Not Legal For Use In Trade", 10 in. long	$23-28
Brass-faced scale, with large iron spring between scale and hook, 13 in. long	$35-40
Brass-faced scale, 120 pound capacity, with hook, 16 in. long	$28-35
Cast iron, with brass scale bar, weights to 300 pounds, "The Plantation Scales/ARC Scale Co.", 28 in. high	$150-200
Scale, balance, brass trip, trays, 13½ in. long	$55-75
Scale, balance, polished brass, 11 in. arm	$35-50

Steelyard scale, wrought-iron, stamped "P. William & Co.", 28 in. long.................................. $35-45

Steelyard scale, iron, 18 in. long.................... $25-30

Round scale, metal case with brass face, 12 in. long $25-30

Spring hanging scale, "Chatillon's Improved", weights up to 50 pounds, 13 in. long $20-25

Steelyard scale, "Whitmore/52", hooks and weight complete $24-28

Steelyard scale, hooks and weight, 25 in. long........ $27-32

Scale weight, "C. I. Ball", held by thumbscrew, 4 in. diameter $8-10

Weights, nickle-plated brass, 100 to 4000 mg., fitted wooden box with advertising by Whiting Machine Works, Mass. $32-37

Weight scale, brass arm graduated to ounces, illegible label, tin platform $30-35

Scale catalog, by E. & T. Fairbanks & Co., Johnsbury, Vermont; printed 1854, with 54 pages describing and picturing the Fairbanks scale line, woodblock illustrations $40-45

Scrapers

With controlled-depth blade edges, scrapers are smaller, somewhat handier, versions of woodworking planes. Typical-width blades are 2 inches wide, and some have long wooden handles. In some types, the blade is mounted in a swivel head so it can cut with both a pushing and pulling motion.

Scrapers could not replace "shaving" planes, for their work is often lighter, and they were used with somewhat more care. Scrapers are aptly named, as most removed only a thin shaving of wood.

Values

Scraper, adjustable, "Bennett Specialty Mfg. Co./ 1-5-09", 11½ in. long $22-26

Scraper, handled, triangular blade, "Vaughan & Bushnell Mfg. Co.", 17½ in. long $20-23

Scraper, "Stanley #80", good blade, 12 in. long $25-30

Scraper, "Stanley #81", mahogany sole, 10½ in. long .. $36-40

Scraper, 'Stanley #151", marked blade, 10 in. long $21-25

Scraper, "Stanley SW", brass sleeve, walnut handle stamp-dated "1898", 13 in. long................. $38-45

Scythes
(See also Grain Cradles)

The long-handled scythe was an omnipresent farm tool, and it had several uses. Not generally recognized is that there were at least three scythe varieties, each made for a different task. These might be termed light-, medium-, and heavy-duty in design and function.

The grass-scythe (lightweight) had a long (averaging 30 inches) slightly curved blade, and was used for cutting areas of tall grass. Such a blade would be used where larger equipment could not go, as in fencecorners and on steep hillsides. In use for hundreds of years, very early scythes of this type were used for grain harvesting and hay-cutting. Such scythes are shown in European paintings and woodcuts.

The weed-scythe had a shorter (averaging about 24 inches) and stronger blade, a bit more sharply curved. It was used for heavier tasks, cutting down thick weeds and light brush. These generally used the same wooden handle or "snath" as the grass scythes.

The brush or bush-scythe, however, was almost another tool. The heavyweight blade was well-curved, averaged about 20 inches in length, and was employed much as people use the unique bush-hook today. The bush-hook in fact has been used since before 1900, and the blade would be the ultimate evolution of the bush-scythe, though used with a short, straight handle.

The bush-scythe was used to slash down undergrowth and small trees, and was essentially a land-clearing tool. It used an extra-strong snath, which, like the others, was steam-bent to shape.

Values

Scythe-head, 28 in. long, rusted, maker-marked	$2-3
Scythe, with handle, 31-inch blade, all parts present	$15-20
Scythe, heavy-duty blade 21 in. long, good handle	$12-17
Scythe, brush-type, well-worn blade, heavy handle	$9-14

Seeders

Seeding non-row crops like alfalfa and timothy hay was done on already prepared soil, usually disced and harrowed. Very small seeds were distributed at first by hand, thrown from a shoulder-slung bag that greatly resembled that used by newspaper carriers.

Of cotton or linen, the bulging bag gave rise to the present media term, broadcast. The seeds were thrown out at random, over a wide area. Something was needed to give a regulated seed-flow, and more careful distribution.

Factory-made devices soon appeared, of two practical and simple types. Each had a cloth seed bag above the mechanism; the seeds fell through regulating slots and struck a spinning, ridged wheel of tin or light iron. This distributed seeds left, right and ahead. The farmer then had uniform seed-fall and some control over seedling density.

These mechanical seeders propelled the seeding disc in two ways. The more common type had a side hand-crank with a wooden handle, and the sower turned this at a steady rate. More unusual was the seeder with a long, wooden bow-stick. This was pulled to and fro to activate the seeding disc. Seeders in good condition are prized.

Values

Broadcast bag, linen, leather strap and lip, very worn . . $12-16

Seeder, crank-action, torn bag, weathered, mouse
    damage $18-25

Seeder, well-used condition, crank-action, original
    maker's mark $40-50

Seeder, bow-stick action, maker-marked, like-new condi-
    tion, original label and instructions on underside . . $80-95

Sharpening Stones

A large class of sharpening devices was used on farms of all periods and places. Simply, they were needed to keep the cutting edges of bladed tools keen and efficient. From axe to chisel and plane to scythe, a dull edge meant a near-useless tool.

In early years, almost any loose-grained rock could be, and was, used to sharpen iron and steel edges. The belt of Berea sandstone stretching south of Lake Erie was a favorite material, and many of the larger mounted grindstone were made from it.

Hand-size sharpeners are often called whetstones, and early

cradle- and scythe wielders carried stones in special leather or wooden sheaths. Few farmers could accomplish brush-clearing or harvesting without a favorite stone in belt or pocket.

Eventually, when more products were available and buyers more discriminating, oilstones became popular. These were small-grained stones with high abrasive qualities, using light oil as a lubricant. Arkansas stone was a favorite, and could be purchased in coarse, medium and fine honing grades. Often such valuable stones were packaged in rectangular wooden boxes which gave permanent protection to prevent breakage.

Another type of sharpening stone, used especially with sickle and scythe blades, was round, about 7 inches long, and pointed at each end. Stroked along the blade edges, it involved careful turning to whet the metal evenly.

Scythe-sharpening by hand created a rhythmic, musical sound that carried some distance. Beyond the smaller, freehand stones, three other sharpening stones deserve mention. One is the mounted "shop" type, a wheel 4 to 10 inches high, with a simple hand-crank. These were secured to work-benches in the tool shed or repair area.

Perhaps the best-known large farm stone was the platform grindstone. "Keeping one's nose to the grindstone" once meant an industrious worker that took good care of tools and business. This was a large wheel mounted on a wooden frame, with the user facing the wheel, perched on a bicycle-like seat.

He propelled the turning wheel by pumping one or two wooden foot pedals, each attached with metal arms to side-wheel levers. These were common on late-1800's farms, and could sharpen nearly any tool the farmer possessed. The wheel was kept cool by immersion of the lower portion in a water reservoir, or water dripped from a pole-mounted can over the wheel.

Patent sharpeners came next. Since the scythe blade was one of the more important long cutting edges, several factory-made types appeared. Unusual today, they were plank-mounted, and crank-operated. Screws held the blade firmly, at the proper angle, while a system of sockets and joints moved the revolving grindstone along the blade edge.

Later came compositions substances of harder material and power tools that nearly made the stones old-fashioned. Sharpening stones have not received the attention they deserve. They were, after all, the tools that permitted the use of other tools.

Values

Rectangular whetstone, slightly used	$1-2
Oval whetstone, never used	$1.5-3
Early stone in leather belt sheath, well-used	$12-15
Duo-pointed round scythe sharpener, dark stone, 7 in. long	$5-9
Flat sharpening stone, light brown, 6 in. long	$2-4
Arkansas oilstone, 2x5 in., marked wood case, lid	$8-11
Arkansas oilstone, 3x7 in., wood case and lid	$10-13
Platform and seat grindstone, poor condition, missing pieces, 30 in. high	$15-25
Platform grindstone, fine condition, all orig. parts, metal seat	$55-80
Sharpener, bench-mounted, hand-cranked, screw-turn base mount, old	$12-15
Patent scythe sharpeners, depending on design, condition and operation method, each	$40-75

Shoemaking and Repair
(See also Harness Making)

While not all early farmers had the skill or spare time to actually make shoes from varied thicknesses of leather, most could and did make frequent repairs. While women might spin or weave on a long winter night, men were likely perched in a corner, repairing shoes and boots.

Not every farm had the professional's cobbler's bench, with low seat, form tree and raised compartment and drawers. In at least one case, the shoe repair gear was kept in an old wooden salt-herring box. Forms or lasts were, early, of carved wood and many have the sole area quite splintered from the countless shoe nails pounded in.

More often seen are the numerous cast-iron lasts and supports, these in size from toddler to adults. The forms hinged or dropped on the support top and the base was securely nailed or screwed down to a surface or heavy base.

Shoe or boot-making required following patterns and measurements to properly size the result, blending tops and sides with sole and heel. Leather-cutters of many kinds were used, and the leather-working pliers was ever-present.

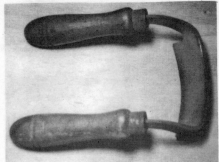

Scorp or concave two-handled scraper for scoop-work, hardwood turned handles, brass sleeves, 8 in. long.

$25-30

Scythe, two-handled snath, well-worn blade 30 in. long.

$20-25

PUBLIC SALE

The undersigned will sell at Public Sale on the Abraham Spitler farm one and one-half miles east of Pleasantville, on the Somerset road

Thurs. October 6, 1910

THE FOLLOWING PROPERTY TO WIT:

1 Gray Mare ten years old, broke for a lady to drive

6   HEAD OF CATTLE   6

Consisting of 3 Milch Cows, 1 just fresh, 1 to be fresh in November and 1 giving milk. 1 yearling steer will weigh 1150 lbs 1 spring steer calf and 1 heifer calf.

Road Wagon, Buggy, Phaeton, 2 sets Single Harness, Sleigh, 300 shocks of corn to be sold in shock and 20 bu. of potatoes.

HOUSEHOLD and KITCHEN FURNITURE

Gas Range, Wood Cooking Stove, 2 Gas King heaters, 3 Breakfast Tables, Extension Table, 2 Safes Cupboard, Flour Chest, Bureau, 2 Dressers, Wardrobe, Chest, Sewing Machine, 5 Stands, a lot of Chairs, Washing Machine and Wringer, 20 gallon Copper Kettle, 2 Brass Kettles, 3 Iron Kettles and a lot of other articles too numerous to mention.

Farm for Sale

The Abraham Spitler farm will be sold at Public Auction on day of sale at 2 o'clock, consisting of 112.80 acres. This farm is one of the best in the county, on a pike road, two good houses, good barn, corn cribs, granaries and buggy house all in good condition. Terms will be made known on day of sale.

Sale to Commence at 10:00 O'clock, A. M.

Terms: A credit of six months will be given with approved security on all sums over $5.00. Under $5.00 cash on day of sale. 2 per cent. off for cash. No property to be removed until settled for.

T. J. SPITLER, Auctioneer.
R. E. SIVES, Clerk.

T. J. SPITLER, Agent.

Sale bill, 1910, framed; for-sale items include: "Road Wagon, Buggy, Phaeton, 2 sets Single Harness, Sleigh, 300 shocks of corn to be sold in shock and 20 bu. potatoes."

$25-30

191

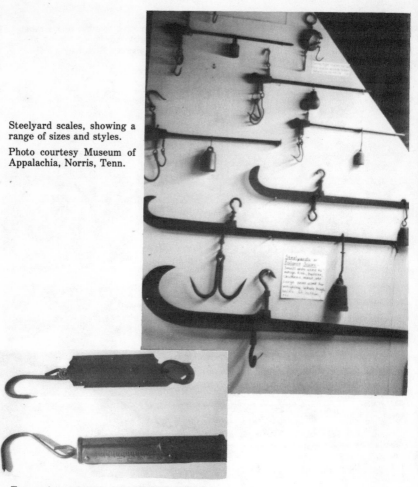

Steelyard scales, showing a range of sizes and styles.

Photo courtesy Museum of Appalachia, Norris, Tenn.

Top, spring scale, 8 in. long including hook and eye, iron face, marked "Pocket Balance".
$13-16

Bottom, spring scale, brass face, iron cylinder body, unmarked, 10 in. long. $16-20

Cobbler's hammer, 7 in. handle.
$5-7

An interesting early collectible is the plank with an elongated, rounded depression, used for pounding out the curved "upper" for footgear. Punches made the holes and heavy needles with waxed thread were used for sewing. The stitching awl was also widely used.

A small hammer was widely employed for all tacking purposes. It had a short handle and a wide, flat head, and was rarely more than 7 inches long. An advantage of the iron last was that it turned and helped flatten nail-heads. Long-nosed pliers were also so-used, but special long-handled tools were needed for boots.

Puncture wounds to the feet from protruding nail-tips could be a serious problem, so great care was taken to rub down or cut off dangerous nails. One long object was the nail-rasp, with small ridges for filing-off tips. Large nails were simply cut off using a similar device with a knife edge that cut when pulled toward the user.

Generally, early shoes and boots were nailed, later types held by stitched thread. In either case, the thick heel was nailed on. Sometimes, for winter work on ice, the farmer twisted in metal studs or caulks for nonslip footage.

Further wooden forms were required for boot-making, these to approximate the lower leg to enhance final shaping. Interestingly, until about the time of the Civil War, boots were not made in "left" or "right" styles, but were straight-toed, to fit either foot.

Values

Cobbler's bench, round leather seat, hardwood legs, five nail compartments, seat area 16 in. high	$150-175
Hammer, shoemaking and repair, 8½ in. long	$5-8
Awl, punching, curved tip, 4 in. long	$3-4
Awl, stitching, with some thread on handle-spool	$5-8
Nail-box, cardboard, with shoe tacks inside, 2 in. long	$1-1.5
Cobbler's light, unusual for a farm, early, consisting of four glass water-filled globes hung on iron frames around a candlestand, 18 inches high	$200-225
Nail-puller, long-nosed plier, 5 in. long	$4-6
Nail-cutter, walnut handle, 15 in. long	$8-11
Nail-rasp, wood handle, brass sleeve, cast iron	$7-10
Cutting tool, crescent, "Newark/1826", 7 in. wide	$75-95

Nail-holder, cast-iron, top 10 in. diameter, compartmented $30-40
Shoe-sole cutter form, cast iron, two pounds weight,
10½ in. long.................................. $17-22

Shooting Accoutrements
(See also Powderhorns and Bullet Moulds)

There were few farm households in the 1700's and 1800's until about 1875 that did not have a lightweight small-bore muzzle-loading rifle. Many were of the familiar "Kentucky" type, with long barrel and perhaps an ornate patch-box in the buttstock.

The under-barrel stock in early examples went to the muzzle, and such scarce examples are called full-stocks. If the stock went part way, they are termed half-stocks. Early rifles used a flintlock firing mechanism, whereby a flint struck steel, producing sparks over the firing pan. Later types used an improvement, the metallic explosive cap, less complex and more dependable in damp weather.

Whatever the basic rifle, the piece was almost useless without a wide range of shooting accessories or accoutrements. An individual rural rifleman of 150 years ago would not necessarily have used every accessory mentioned here. Some, howevever, like bullet patches, were in universal use and all such rifles required a ramrod.

Often lead balls were carried in a separate, small bag called a bullet pouch to keep them together and clean. The pouch might be of leather or linen and contained a few dozen round balls.

The bullet starter was also called a "straight starter" or "ball seater". This was a simple wooden tool with a dowel-like projection about 3 inches long, connected to a round wood handle. Struck with the palm of the hand, it pushed the patched ball several inches down the bore, making ramrod use easier.

Shaped like a miniature corkscrew, the metal bullet worm or "wormer" or extractor was fastened to the ramrod tip. It was a way of unloading without firing. and it "pulled" or unseated an already loaded ball. It was especially valuable when the main charge repeatedly failed to ignite and had to be replaced.

Cap dispensers were known also as "cappers" or "cap magazines". These were metal devices that stored a supply of

Rifleman's bag and horn, well-made example, horn 12 in. long secured to shoulder straps.

$55-70

Leather-working pliers and hammer, 8 in. long, "R. Timmons & Sons".

$7-10

Above
Leather-working awls, longest 6 in. long. Each . . . $4-6

Cobbler's bench and leather-working tools and supplies. Photo courtesy Museum of Appalachia, Norris, Tenn.

Shooting accoutrements; three lidded containers for three different types of metallic percussion caps. Brass boxes, Two, "Union Metallic Cartridge Co.", each . . . $3-4
One, "Goldmark's Percussion Caps", mfg. by "Winchester Repeating Arms Co.". $4-5

Shooting aids; metal cap container and linen cap pouch. Container has sheet-brass base, 1½ in. diameter.

$4-7

Cap pouch, whittled spout, 7 in. long. $8-11

Shooting accoutrements; center, bullet-box with early round lead balls.

$4-6

Bottom center, two-cavity cast-iron bullet mould, 4 in. long. $19-24

Left and right, two Winchester-marked bullet moulds, walnut handles with brass at each end, fine condition. Each . . . $33-40

fresh, unexploded caps. Often spring-wound or driven, they also automatically fed single caps for the firing nipple or cone in caplocks. Only a discriminating shooter carried the factory-made dispenser, as the average rifleman carried caps in the tin box they were purchased in.

Such small, round containers usually held 100 caps. A few shooters stored their caps in small horns, shaking them out the open small end.

The drum wrench was a small tool used to unturn the drum, the metal protrusion which lead through the barrel base side, into the firing chamber. This had to be done occasionally for repairs.

The hunting bag was simply a shoulder-slung pouch, often with the associated powderhorn, used to carry shooting accoutrements. Early bags and horns are more scarce than the rifles themselves. A typical bag might measure 10x14 inches, plus the long leather strap. The hunting bag should not be confused with the game bag, used to carry downed rabbits and squirrels.

The loading block was an efficient time-saver, in that it held prepatched round lead balls. A typical size was ¾x2x6 inches, with perhaps a leather thong at one end for easier handling.

It contained six or eight drilled holes, each which held one ball with patching material. In use, the block was held over the muzzle, one chamber aligned with the bore. The load was then pushed in with the starter or ramrod tip. Most riflemen carried at least one such block for rapid shooting.

For best results and easier loading, a lubricant was needed. Patches were oiled or greased on the side opposite the ball, making them easier to insert. Various products were employed, including a cake of beeswax, lard, or bear grease. Some shooters, when they could get it, preferred pure sperm-whale oil.

Nipple primers are very rare, and the metal devices were as a rule made abroad. Small tubes with plungers, they forced extra-fine powder into the cone or nipple vent and channel. Such extra ignition powder, it was hoped, would explode a troublesome hangfire or misfire of the main powder charge.

A small metal tool, with "L" or "T"-shaped handle, the nipple wrench was used for unscrewing and removing a damaged caplock nipple. This in turn was attached to the barrel drum.

Patches served several important purposes. A patched ball slid down the bore more easily, as the lead sides did not actually touch the rifling. The patch fabric also helped remove powder residue from previous shots, helping keep the bore clean.

Most directly, the patch acted as a seal to hold back explosive gases when the rifle was fired. This increased the velocity of the ball. So a patched ball shot "harder" and "flatter" (with less gravity-drop) than an unpatched ball. Patches were cut from a variety of thin fabrics, and bed-ticking and linen were highly regarded for the purpose.

When quantities of patches were required, the patch-cutter was used. It had a flat top and a sharp cutting edge at the opposite and which was placed over layers of patching material. A mallet-blow produced a number of patches. These were typically round, but other shapes included squared crosses.

The patch-knife was used for sizing patches as the rifle was being loaded. This small-bladed tool cut off any excess patching. The ball was pressed into the bore at the muzzle by thumb, and the blade then trimmed away extra fabric. This aided accuracy since the ball was shot with the same amount of patch each time. Of course, pre-cut patches did not usually need to be sized.

The powder measure was a tube of bone, horn or antler. It transferred the proper amount of powder from the horn to the muzzle. Similar powder charges meant the piece fired about the same each shot, and could not be dangerously overloaded. An early rule-of-thumb stated that the proper powder amount, when poured over the ball, should just completely cover it, sides and top.

Usually made of hickory wood, the ramrod was carried beneath the barrel when not in use. Slightly longer than bore length and smaller than bore diameter, it was used to tamp down the bullet or ball. Light final taps seated the ball snugly atop the black-powder charge.

The rifleman's knife somewhat resembled a medium-size hunting knife. It had a single edge, and many rural shooters carried one at the belt or on the hunting bag. Multi-purpose, they were used for everything from cutting patches to skinning small animals.

The vent-cleaner is also sometimes called a "priming wire". The needle-like instrument was used to clean the firing channel inside the caplock nipple or cone.

After wild game began to be "shot out" and disappear, competitions arose like turkey shoots and the unusual beef shoot.

In the turkey shoot, the goal was to win the turkey by removing the distant, bobbing head with one shot.

In the perhaps more civilized sport of beef shoots, the prizes were beef quarters. Targets were made by charring one side of a board black and the firing marks were traditionally either a cross scratched in with a knife tip or a white center cut from a playing card.

First prize was not a choice cut of beef, but the backing stump against which the target board had leaned. It contained all the lead balls shot by the competition, so the thrifty winner acquired a good supply of lead.

Wherever and however muzzle-loading rifles were used, associated items assured that the pieces would function safely and well. Such shooting accoutrements were once a necessary part of rustic America.

Values

Bullet pouch, cloth, lead balls, 5 in. long	$4-6
Bullet starter, handle 2½ in. diameter, 2 in. projection	$7-10
Bullet worm, double-twist heavy wires, 1½ in. long	$4-6
Cap dispenser, tubular, 4 in. long	$35-45
Cap dispenser, circular, spring-driven, 3 in. diameter	$65-80
Cap box, factory, tin, "Remington", green logo, caps remaining	$5-7
Cap horn, (like miniature powderhorn), 4 in. long	$12-16
Hunting bag, 9x13 in., strap, some shooting accoutrements inside, plus orig. powderhorn	$85-100
Loading block, old, 2x5 in., eight holes	$12-15
Lubricant, grease contained in old cap box	$1-2
Nipple primer, 3¾ in. long, prob. French origin	$125-175
Nipple wrench, "L"-shaped handle, 3 in. long	$3-5
Patches, in ornate homemade wood box, 2x4 in.	$9-12
Patch-cutter, circular cutting edge, 5 in. long	$12-15
Patch-knife, antler handle, 3 in. curved blade	$25-30
Powder measure, empty .38 cartridge, string attached to measure base	$1-2
Powder measure, carved bone, hollow, terminates in animal head, 3 in. long	$35-45
Ramrod, hickory, 37 in. long, plain, hand-whittled	$2-4

Rifleman's knife, stained stag handle, brass guard,
   "Sheffield/England", 6 in. blade $65-85
Vent-cleaner, in screw-on metal case, 3 in. long $9-12

Shovels

As with many other farm tools, shovels were quite specialized.
Some farms had half a dozen kinds, each for certain tasks, and
the shovel category discussed here is extended to include scoops
and spades. It might be noted that most shovels, except for grain
scoops and certain spades, were available with both long or short
handles. The short handle almost always had a reinforced handle
grip, the so-called "D"-shape.

Grain scoops in early days were all-wood, sometimes hand-
carved from a single piece of wood. Most were wide and deep,
and on average heaped full might equal half a bushel. Coal
scoops or shovels were also wide but shallow, strongly
constructed to withstand long wear. Both scoops could be, and
were, used as snow shovels.

The standard farm shovel had a slightly pointed tip and a
somewhat heart-shaped blade; these were used for all general
digging purposes, from root cellar construction to draining a low-
lying field. Again, these came in different sizes, and had either
long or short wooden handles.

The spade family has long, narrow blades. A general types was
used for spading the garden, with a blade about a foot long. The
drain-digging spade, with a blade up to 18 in. long, had rounded
front-edge corners. A near look-alike was the posthole spade,
with square corners on the cutting edge. In the former case, the
rounded edge fitted the contour of the laid-down tiles. The
shovel-like drain cleaner had a semicircular long head that tilted
at an angle. It could be operated via the long handle and used
with either a pushing or pulling motion.

Still another farm-related shovel is the lightweight, long-
handled scoop type with heavy wireware basket and connected
and lipped shoveling edge. These appear at the antiques shows
with various tags affixed, identifying them as everything from
potato and vegetable shovels to coal scoops and live-ember
carriers.

While they were undoubtedly used for many purposes, some
original labels refer to them as coal shovels. Probably, they were
popular for tasks like loading the smaller cast-iron stoves.

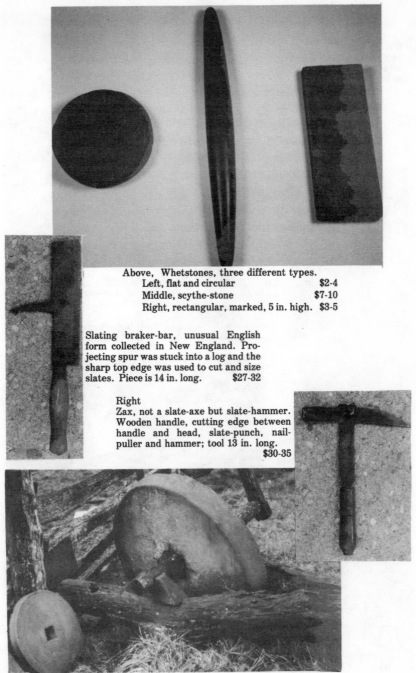

Above, Whetstones, three different types.
Left, flat and circular $2-4
Middle, scythe-stone $7-10
Right, rectangular, marked, 5 in. high. $3-5

Slating braker-bar, unusual English form collected in New England. Projecting spur was stuck into a log and the sharp top edge was used to cut and size slates. Piece is 14 in. long. $27-32

Right
Zax, not a slate-axe but slate-hammer. Wooden handle, cutting edge between handle and head, slate-punch, nail-puller and hammer; tool 13 in. long.
$30-35

Grindstone, large size, with interesting mounting in crotch of tree; hand-powered with side crank.

Photo courtesy Museum of Appalachia, Norris, Tennessee

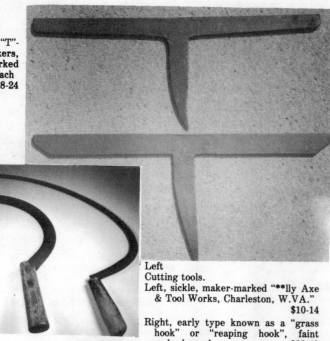

Slating tools, "T"-bars or breakers, both maker-marked on top-piece. Each $18-24

Left
Cutting tools.
Left, sickle, maker-marked "**lly Axe & Tool Works, Charleston, W.VA." $10-14
Right, early type known as a "grass hook" or "reaping hook", faint maker's mark. $30-40

Large frame-mounted grindstone of Berea sandstone, about 30 in. high overall. $30-40

Slater's artwork, near Lancaster, Ohio; note, barn siding has since been replaced. Full reading, which surrounds the primitive-style horse is: "T. B. Reese / By S. H. Vagnier / 1888".

Lar Hothem photo

202

Values

Scoop-shovel, primitive, short "D"-type handle, all wood	$75-90
Scoop, steel, short wood handle, metal-strapped handle	$9-12
Scoop, coal, low straight sides, square leading edge	$6-9
Shovel, rounded tip, rolled-steel top edge, long handle	$8-11
Spade, garden, squared edges, "D"-handle top	$7-10
Spade, drain-digging, 18 in. blade length	$10-13
Spade, posthole, squared blade edge, short handle	$11-14
Drain-cleaner, 6-ft. handle, swivel-and-catch head	$10-13
Coal shovel, long wood handle, wire basket scoop	$23-28

Sickles

Sickles or "grass hooks" were widely used in fields and gardens, especially in clearing weeds from around posts and buildings. This smaller relative of the two-handed scythe had a curved, rounded flat blade, with the whole looking something like the top of a question mark. They were made in small and medium sizes, and most had a squarish or round wooden handle.

A similar type, for one-hand use, was the "scythe hook". It had a slightly curved blade strongly back-ridged. The handle was up-angled to allow free use without the hand striking the ground. Though still a type of sickle, the blade more resembled, that of the scythe.

Both sickle types were sometimes made with gently serrated edges. These were designed to give greater weed-cutting ability, but such edges were difficult to resharpen. Though either could have been used for harvesting, they were developed later and were used more for weed and undergrowth supression.

Not so their ancestor, known as the "grain hook" or "reaping hook". This was a very long, deeply curved blade, triangular in cross-section, with a small wooden handle. Most were hand-wrought, occasionally maker-marked, and very graceful in outline.

The blade ended in a fairly sharp tip, often somewhat blunted from strikes against the ground. Reaping hooks were used with grain rakes for collecting, then hand-tieing, the grain bundles.

Values

Sickle, grass-hook, factory-made $7-10

Sickle, scythe-hook, factory-made, serrated blade $8-11

Reaping hook, wrought-iron, tanged blade in fruitwood
handle $25-35

Reaping hook, wrought-iron, maker-marked, extra-long
blade $30-40

Slating Tools

Today, with the availability of synthetic roofing materials of many kinds, it is hard to recall the importance of slate roofs. At one time, most well-built farm edifaces were complete only when the overlapped slate shingles had been put down. While almost all brick buildings had slate roofs, so did many of wood construction. Though professional slate-layers existed, the average farmer could also do some of his own work.

Given the complexity of the task—laying slates so as to be both watertight and eye-pleasing—the slater's tools are surprisingly few. There are only three main types, though varieties exist. And, since not many people recognize the scarce tools, bargains can sometimes be found.

The most important tool is the zax, a four-in-one tool. On most examples the functions are obvious. Handles are about a foot long, and the curved head is typically about 9 inches high. The handle ends are sometimes of hardwood or compressed leather washers. The leather-handled piece might be a factory product, some other types handmade. The remainder of the handle is steel, and it and the head are generally cast or wrought into one unit.

One end of the zax head is a hammer, while the opposite terminates in a pick-like point. This made holes in the slate shingle, while the hammer drove in the roofing nails. And, between the two, where the handle joins the head, is an offset nail-puller for extracting old or misplaced nails. These have one set of claws, sometimes two, large and small.

To complete the all-purpose function, the metal handle portion has one, or both, edges beveled to form a long cutting edge. Thus, the zax could be used to trim or size, then set, the slates.

Used in conjunction with the zax is a metal device called the stake, breaker or T-bar. This consists of a steel strip about 1½ inches wide, ¼ inch thick, and 18 inches long. Projecting from one side is a long, slightly curved flange with a pointed end.

In use, the T-bar was driven into a stump (if used on the ground) or a roof timber if the slates were trimmed just before placement. The slate was held over the T-bar or breaker cutting edge at the appropriate line and the zax was used with a downward motion to cut the slate. For all practical purposes, the zax edge and T-bar edge formed a sort of gigantic shears, and it took some skill to make clean cuts.

Perhaps the scarcest slating tool—and there is a reason for this—is called a ripper. All metal, about 2 feet long, it has an offset handle and a long extension about 2 inches wide. the working part of the ripper is an enlarged sharpened end with an indentation in the center and hook-like flat metal fingers back-curved on each side of the head.

The ripper (sometimes called a "shingle-thief") was used to replace broken or damaged slates. The end was pushed under the slate to be replaced, and the indentations settled against a nail. A sharp hammer-blow on the handle base cut the nail. The hooks pulled out nail pieces so the new slate could be properly inserted and settled.

Rippers are probably hard to find because few were ever made, since few were used. That is, once a slate roof had been correctly installed, it hardly ever needed repairs. For example the roof of the writer's 1872 brick farmhouse, while containing some broken shingles, shows no evidence of major repairs or replacements.

The last category of slating tools combines several of the other tasks. These are the platform-mounted "machines" known as slate-cutters. Screwed or nailed onto a plank, it operated like a huge office paper-cutter. A long hinged bar fitted into a groove between two straight ridges. The slate was positioned so that the descending cutter took off the required amount of slate.

Most such cutters had a punch-like pin that could also hole the slate. Such cutters seem to have appeared well after the advent of the zax and T-bar. They were popular because they were easy to use and gave better control than the "freehand" methods. While cutters do not appear to be especially sturdy, at least one specimen has a frame of cast brass.

In the Midwest, and other parts, it was somewhat of a rural tradition for the farmer to have his name worked into the house or barn slates, as well as the date the building was constructed. Many examples can still be seen, most dating in the 1880's and -90's.

It should be noted that slating tools are still being made today, and are used for installing modern slate roofs. Old-time workers would have whistled in amazement at the price for a custom-laid slate roof. Pennsylvania black slate can cost over $150 to cover 100 square feet of roof, while New York red slate can go over $1000 for the same area.

Values

Zax, 9¾ in. head, side nail-puller, wood handle	$22.5-27.5
Zax, leather-washer handle, single cutter edge	$25-30
Zax, turned wood handle, oblong, factory-cast, marked	$25-32.5
Zax, hand-wrought, early, 12½ in. long	$15-20
Breaker bar, 17 in. long, early weld, battered	$10-15
Breaker bar, maker-marked, like-new condition, 18 in. long	$20-25
Breaker, bar, marked, 18¼ in. long, good steel	$17.5-25
Ripper or shingle-thief, handwrought, 19 in. long, rough	$15-20
Ripper, factory product, 25 in. long, symmetrical, iron .	$30-40
Slate-cutter, beam-mounted, steel frame, maker's mark, 23 in. long, good condition	$45-55
Slate-ripper, "Keystone Saw Works/Philadelphia", 25 in. long	$35-45

Soap-Making

Early farm people made soap periodically, and it was no small task. All the ingredients could be easily obtained, beginning with caustic potash or lye. Potash (potassium hydroxide) as the name suggests, was obtained from wood ashes. Boiling water was poured through the ashes and the liquid run off as lye-water.

This was evaporated and mixed with skimmed and filtered fat, usually in the form of bacon grease. Eventually, cooked together, a thick, gruel-like substance was formed. The liquid soap was then poured into trays or moulds to harden as rough-and-ready lye soap. Sometimes an effort was made to scent the potent cleanser with rose petals or sassafras bark.

Soap-making collectibles may have to be identified by an older family member, who could remember the item actually being used for the purpose. Items include kettles, stirrers, filters, ladles, trays and block-soap cutters, as well as soap-boxes and dishes.

Values

Dipper, treen, bowl area leached off-white, 17 in.
    long $60-75

Tray, for hardening lye-soap, wood plank construction,
    3x14½x20 inches $25-32

Stirrer, soap-making, flat end leached white, 21 in.
    long $7-10

Kettle, iron, bailed, 15 in. high, reserved for lye-soap . . $35-45

Soap cutter, iron, knife without handle, made from corn-knife, hand-forged, 15½ in. long $20-30

Soap box, wood, wax-paper bottom liner, dovetailed
    corners, no top, 13x15x21¾ inches $30-35

Soap dish, wood, rectangular, 3¾x7½ inches $33-40

Funnel, turned wood, for filtering lye-water, 13 in.
    high $40-50

Spears, Gigs, Etc.
(See Fishing Decoys)

Nearly every large older farm, as one can still see, had a stream on the land or at least nearby. While we think of obtaining freshwater foods in terms of sport, people once fished for the more basic purpose of diet supplement. When this harvest of wildlife took place, the flesh was either eaten in season or perhaps smoked or dried for coming lean months. Such activities were more important than is generally realized today, as the number of related objects is still large.

Probably the most common farm fishing collectibles are the iron fish-spear heads. Earliest examples were hand-forged by expert iron-mongers, either by the local blacksmith or a farmer with special skills.

Many examples had four projecting tines, each with a barbed tip or sides to catch and hold the fish. While lightweight and delicately worked examples can be found, most of this type are about 5 in. wide and 6 or 7 in. long not counting the haft or manner of handle attachment.

The handle arrangement could be socketed to accept a pointed wooden pole or be "rattail" to insert into the center of the shaft end. Most are well-balanced, and show excellent symmetry.

By the late 1800's, factory-made examples came into wide use, evidenced by a ledger entry from a Pennsylvania farmers' supply store: "1 harpoon, by cash, 45 cents". Likely, this was a fishspear head, similar to those sold by Sears and Roebuck in 1908. Theirs were five-pronged, at 41 cents each. Factory-made heads were usually forged or cast, and lack the squared configuration of handwrought types.

Fish-spears were usually sold without the handle, though several very scarce examples have an iron handle shaft and wooden end. It is unusual to find a good head with the original, or at least old, long hardwood shaft.

Some genuine harpoons exist (these heads having several strong barbs) designed to detach from the handle and be held by a strong cord. Some early accounts mention they were also used to capture fur-bearing animals. The collector will be lucky to find one.

Frog-gigs are much like fish-spear heads, except they are both smaller and lighter. Many have socketed or screw-in hafts, have from 3 to 5 tines, and may have been japanned in green or black glossy paint. Most were not strongly made and not many exist today.

Eel-spears are another category, and many antiques dealers call fish-spears by this name. However, unlike fish-spears, genuine eel-spears were wide and flat, and not really designed to impale the slippery prey. Instead, tines were rear-hooked, to slide around the eel and catch it when the head was jerked sharply backward, toward the user. Fine examples can still be found at some New England rural auctions.

Eel-traps can occasionally be seen, sometimes a bit misshapen from long immersion in water. The old types are of woven wood-splint, conical, and about 20-24 in. long. They were designed so that eels swam into the larger end to be trapped in the dead-end.

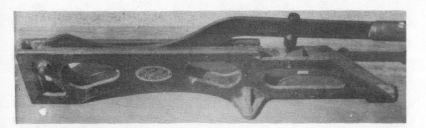

Slate-sizing and punching machine, lever-operated, 18 in. long. Marked "Parsons Bros. Slate Co. / Pen Argyl, PA / SLATE - The Best Roof In The World". Body is cast brass, with black japanning.
$40-50

Try-squares, three different types, with measuring steel blade in foreground 6 in. long. All have a decorative brass inlay at the join, and a brass bolster-strip on the front of the wooden section. Each... $12-17

T-bevel, adjustable measuring face, brass handle ends and screw, exotic hardwood handle. Item, folded, is 7 in. long. $12-15

Below
Try-square, with wood handle brass-bound on two sides, The solid brass device shown with it is said to have been an arc-maker, used to determine angles maller than 90 degrees.

Try-square	$12-17
Brass arc-maker	$15-20

Right
Still, homemade of copper, showing cooker (background) and distillation coil (foreground); used to make moonshine whisky.

Photo courtesy Bob Evans Farms, Rio Grande, Ohio

209

Right
Riding aids; top, bentwood stirrup about 6 in. high. $6-8
Bottom, cast-iron spur. $3-4

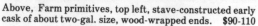

Above, Farm primitives, top left, stave-constructed early
cask of about two-gal. size, wood-wrapped ends. $90-110

Bottom right, cooper's tool for barrels, handmade, wood.
                                                      $45-55

Right
Distillation chamber from farm corn still, all copper with
copper rivets. End is of handcarved wood, from Kentucky, 13 in. high. $15-23

A variety of wooden containers, mostly stave-constructed, but including
solid-wood and bentwood. Item at lower left is a salt container.

Photo courtesy Museum of Appalachia, Norris, Tennessee

Odd-seeming today, crayfish or "crawdads" were once a significant freshwater food, being treated as little, local lobsters. They could be lured from the bottom of ponds, streams and springs by a baited string and taken with a small dipnet.

Some of the least-known devices for obtaining underwater creatures were the so-called "turtle tools". Everyone is aware of turtle soup (tasting somewhat like a beef stew) but a few know how the snappers were caught. Some old-timers swore by "noodling", walking a streambed and reaching into holes in the bank. Others used heavy hooks on multi-stranded trotlines.

More careful folks probed the bottom of ponds and marshes and scooped up the turtles with gear that held the turtle and kept it away from vulnerable fingers. E. F. Robacker, in the book, *Old Stuff in Up-Country Pennsylvania*, shows a locally made "turtle hook". It is a pitchfork with the four tines welded together at the tips to form a long-handled basket.

Another Midwestern example, evidently made from various parts, is both a probe and hook. It is in two sections, connected by a hollow brass sleeve. About 4 ft. long, one end has a 7 in. tip, the other a 3-pronged fork with flattened ends. This piece was made to both locate and capture hidden turtles, and is probably a century old. While not a great work of art, it at least displays native ingenuity.

Values

Hand-forged fish-spear head, 4 tines, socketed haft. . . .	$12-18
Hand-forged fish-spear head, 5 tines, "rattail" haft	$15-20
Hand-forged fish-spear head, factory-made, iron handle 4 ft. long, wooden handle cap	$25-35
Factory-made fish-spear head, 4 tines, early 1900	$8-12
Factory-made fish-spear head, 8 ft. wooden handle	$25-40
Harpoon, head, of cylindrical handmade iron, one barb	$15-20
Frog-gig, socketed haft, 3 tines, factory paint	$4-7
Eel-spear, 4 flat tines, head only, 9 in. long	$30-40
Eel-spear, center spatula, 6 protrusions, broken handle, hemp-wrapped at head base	$45-60
Eel-spear, 7 prongs, 9 ft. handle, from Maine	$55-75
Dipnet, metal handle, woven net, 10 in. diameter	$5-8

Turtle tool, walking-stick length, pointed end $6-10
Turtle tool, pole with probe and metal scoop at opposite
    end.. $15-20

Spurs

When riding-horses were the main locomotion, spurs served a useful purpose. If the reins can be considered the steering wheel of the animal, spurs served as the gas pedal. Metal devices that strapped to boots, they were useful for racing and controlling the animal.

Spurs can be the ornate Southwestern Spanish style, or sturdy as in the Civil War or down-to-earth farm types, simple and solid. Spurs were made of metal, usually iron or steel, but sometimes of cast-brass. Matched spur sets are most in demand, though single specimens can be found. Most sought-after are the hand-tooled examples with precious materials.

Values

Single spur, iron, small plain rowel $4-5
Spurs, pair; brass, orig. fastening chains $23-29
Spurs, pair; large, ornate rowels, silver-mounted $110-130

Squares, Etc.

The square was a flat metal tool useful for both measurements and marking right-angles, and was large enough to cover even wide boards. Most were several feet long, with a projecting side arm known as the tongue. Squares were needed, for example, to measure and mark the straight-edge lines for making a barn-beam tenon or mortise.

Try-squares were much smaller, often with handles of exotic woods like mahogany or rosewood. Joined to the iron measure and held with several pins in decorative brass, the inside wooden handle was usually faced with non-warp brass. Try-squares marked off straight lines at right angles to a board's exterior surface.

The miter-square was similar to the try-square, except that the handle-rule join area did not form a 90-degree angle, but a 45-degree or miter angle. This was ideal for marking and measuring everything from window corner frames to making doors.

A T-bevel again had similar characteristics—wooden handle, metal measuring strip—with several differences. The strip was not rigid, and could be angled in about 270-degrees of arc. Further, the pivoting measure could be fastened and held in one position by a locking lever or thumbscrew at either end of the handle.

Some patent devices combined, with extra attachments, a try-square so that it could make all angles or bevels. It was known as a bevel-square.

Values

Square, 2 ft. with 18 in. tongue, maker-marked $7-10

Try-square, "Stanley/Pat. 12-29-96", brass-strapped . . . $11-15

Miter-square, less common than try-squares, un-
    marked $14-19

T-bevel, rosewood handle, brass locking nut $10-14

Bevel-square, maker-marked, brass reinforcements $18-24

Stave-Constructed

Old containers constructed of wood staves are a collecting field in themselves. Very many separate types exist, small to large. Stave construction was of two kinds, wet (for liquids) and dry (for loose solids). In general, because of the tight and careful fit, wet-stave objects were better done, with high-quality woods.

Values

Bucket, stave-constructed, interwoven bands, wood
    handle, 11¾ in. high $55-75

Bucket, sugar, staved, with lid and handle, 12 in. high . $85-100

Canteen, wood, horizontal staves, 9½ in. high $20-25

Canteen, staved, worn red paint, 9 in. diameter $40-45

Canteen, staved, 6¼ in. diameter $50-65

Cask, whiskey, stave-constructed, metal rims, 4 in.
    diameter $22-35

Cask, staved, hickory handle, 12 in. diameter $70-90

Keg, stave-constructed, 20 in. high $45-55

Keg, stave-constructed, 4¾ in. diameter $17.5-22.5
Measure, wood staves, outer lapped band, 5 in.
    diameter $45-60
Measure, staved, worn red paint, 13 in. high $60-70
Keg, staved, iron-bound, 13 in. high $30-40

Stills
(See also Corn-Related)

In early rural days, from about 1750-1850, corn was plenti-
ful, but freight transportation—whether by land or water—
was scarce and expensive. So, many farmers concentrated their
corn into alcohol, which was much more profitable to ship.

Corn liquor sold both widely and readily and was a major
barter item. For many years, especially in the early 1800's, no
threshing day or barn-raising was complete without a nearby
keg of corn-distilled whiskey.

Besides corn, hard cider could be further treated to make a
kind of potent brandy, so also peaches and other fruits. Stills of
all sizes produced and recovered the alcohol. Some were
makeshift and inefficient, while others were well-made from
non-contaminating materials.

Values

Still, copper boiler, copper distilling coil, 15 gal.
    capacity. $125-150
Still, part, coil only, copper tubing, 39 in. long. $25-35
Still, part, made from copper clothes boiler, two holes for
    tubing $25-35
Still, tin, all parts present, small, early $60-75
Still, brass container, copper tubing coil, buffed and
    lacquer-coated $150-175

Stock Markers
(See also Branding Irons)

Rural crime is not a new thing, and animals were sometimes
stolen. Too, marketing—sending cattle or swine off to distant
selling terminals—presented identification problems. If the
herd was small but being sold miles away, a permanent record

was needed. Also, if the herd was large and needed periodic veterinary treatment, this required some form of ongoing animal recognition.

Stock markers were soon applied to individual animals, though branding stamped the animal as to owner. Four different methods were used. One was the ear-button, a circular aluminum rivet-like addition to heavy ear-parts. Ear-labels—thin metal strips—were also used. Another method was ear-clipping, removing edge-segments with a cutting or notching pliers.

Tatooing was also done on animals from dogs to horses, using a pliers-like pincher with raised lettering and colored ink. Such marks were both tamper-proof and permanent.

Stock tags or "stock-marks" were of several kinds, with cast-iron examples sometimes fastened on with a ringer and metal rings. Solid-copper stock markers were also available; most such tags carried a number or letter.

Values

Ear-button, aluminum $0.75-1.25
Ear-labels, unused, in frayed old cardboard box $3.5-5
Ear-clipper or notcher, triangular jaws, rusty $6-8
Tatooing pliers, inset numbers, raised, 8 in. long $9-12
Stock tag, cast-iron, 1¾ in. high $3-4
Stock tag, heavy copper, with letter designation $5-7

Stoneworking

Working with stone was crucial on farmsteads, and many men and boys became proficient with the unique tools. Shaped and dressed stones were used for outbuilding foundations and corner-stones, and most barn designs would have beenimpossible without major stonework. And the stepping-stone was a necessity for raised doorways.

Hard to locate but fine examples of stoneworking arts are downspout collectors for channeling rainwater away from a building. Stone watering troughs for farm animals were not always made from mortared stone blocks. Some were cut one-piece, from solid rock and moved on rollers to where they were needed. A few stone fenceposts exist, either holed or indented for wooden rails, and they will nearly outlast time itself.

Some backwoods springs were enhanced with stone basins, and these, emerging from hillsides, had vaulted stone ceilings. Such constructions are treasured and are no longer being ignored or destroyed.

Compared with other farm tools, stoneworking examples are fairly scarce. And often, they are not recognized for what they really are. Such tools share certain characteristics. They always have very heavy heads, generally of cast-iron or cast-steel. Edges, when present, may be treated or tempered to extra hardness, and they are always blunt-edged.

Handles tend to be short and sturdy. One variety almost always has a cast-iron handle, while most other handles are hardwood. In a number of cases, wood handles were broken from the heads and not replaced.

Perhaps the most interesting stoneworker's tool—called by some tool collectors a "crandle", with spelling uncertain—has a set of loose teeth in the split head. Often there are 13 long, square and pointed teeth held securely by an iron wedge. There is also a cast-iron handle portion, very heavy, about 12 in. long.

The loose teeth allowed a certain "play" when used, and assured an even strike that left uniform impressions. Also, dull or broken teeth could easily be removed for repair or replacement. Despite the short handle, this tool was used with both hands due to the great weight, about 10 pounds. Rough-stamped owner's names are sometimes on the handle, or a maker's mark.

Another tool type has a heavy duo-bladed head and a short wood handle. The edges on both sides are serrated to provide a number of separate striking surfaces, allowing enormous impact in a relatively small straight-line area. Head weight is about 5 pounds, making this stoneworking tool useful with one or both hands. This tool may be scarce, because while the head was cast, the edge divisions appear to have been handmade.

A number of specialized stoneworking hammers exist, in three main categories. All have long, narrow and fairly wide heads, with an oval handle opening. They are all duo-headed, and are listed here by different striking surfaces. These might be described as, edge/edge, edge/flat, and flat/flat.

The edge/flat type is probably the most common. Sometimes the heads only were purchased and the user made his own handles. The heads had typical weights of from 3½ to 5 pounds. The edge/flat or single-faced was often made of cast-steel and with a polished flat or face.

The flat/flat type, with two opposing square-corner rectangular striking surfaces was called a double-face spalling hammer, and was offered in about the same head weights.

The edge/edge might offer two striking surfaces of slightly different characteristics for broader results. Such hammers were offered in many styles and weights, and are likely the most common stoneworking tools to be found.

Stone sledges had long handles and head weights of 8 to 12 pounds, with one round, flat surface and one edge on the large head. The familiar sledgehammer, however, generally had two flat surfaces, and came in weights of 3 to 9 pounds, with short or long handles.

Smaller drilling or striking hammers were used with stone-punches and chisels. They struck such steel instruments for smaller farm jobs, like removing imperfections or making designs. Hammer and chisel were sometimes used to carve folk art creations.

An important stone-maneuvering and lifting tool was the long and heavy steel bar which had several names depending on the region of the country and tip design. These were up to 5 feet long and weighed from 12 to 20 pounds. Called crowbars, pinchbars or prybars, the handle was typically rounded and the lower portion squared. An even longer bar with a wide, rounded or squared edge was called a spud.

Other devices associated with farm stoneworking include the ways and means of bringing stone to the building site. For smaller stones, there were two-man carrying bars, like that used by loggers, except with a platform instead of tongs.

The stone barrow was good equipment, like a wheelbarrow, but much sturdier, low-slung, and with open sides. Another transport device was the stoneboat, a heavy-duty sledge pulled by several horses.

Accessories to the tools included chalk-lines and other marking pointers. Also valuable were transits and levels, slings and pulleys, squares and plumb-bobs, tool-chests and spare wooden handles. Not a great deal is known about stoneworking tools and those who used them. Perhaps one day an early journal will be found, and it should be fascinating reading.

Values

Hammer, edge-flat, 4 pound head, replacement
    handle $14-18

Hammer, flat-flat, 4 pound head only, handle broken . . $8-11

Hammer, edge-edge, 3½ pound head, old if not original
    handle, 14 in. overall length $15-19

Crandle, all teeth present, maker-stamped on handle,
    6¾ in. teeth $40-50

Crandle, 13 teeth 9½ in. long, lightly used, little rust,
    17½ in. overall length $35-45

Striker, serrated head, straight-edged, short wood
    handle, 4½ pound weight $19-24

Hammer, drilling, duo octagonal-sided heads, short
    handle $7-10

Stone sledge, flat and cutter head, 10 pound weight,
    good old handle $12-15

Sledgehammer, long-handled, 12 pound head, cast-
    steel $10-15

Sledgehammer, short-handled, 4 pound head $7-10

Punch, stoneworking, toothed tip, 11½ in. long $6-9

Chisel, stoneworking, shows much wear, 9 in. long $4-7

Prybar, square chisel tip, round handle, 51 in. long $19-25

Stone barrow, old, wheel collapsed, termite damage . . . $12-18

Chalk-line, stoneworker's, on wooden reel, 10½ in.
    high $11-15

Straw

Grain-stalks and leaves, the straw, were almost as important to the farmer as the grain itself. Ryestraw—the stems of the herb-grain, rye, important in baking and whiskey-making—is an example. The main use for the long, slender straws was basket-making, and ryestraw baskets exist in numbers for the collector in the Ohio-Pennsylvania region.

Ryestraw was gathered and wet down so that the long sections became pliable. They were then gathered and wrapped

Toothed stoneworking tool, teeth not resharpened in this example. Square-bodied teeth are held in with iron wedge; signed "King". $45-60

Serrated-edge stoneworking hammer of unusual design, head signed with indented initials. Head is 9 inches high, original or old wood handle.

$22-30

Spalling hammer-heads, with remains of broken handles. Each is about 7¼ in. high. Top, edge/edge type, bottom, edge/flat type. Each . . . $10-14

Whiskey still, small size, from Claiborne County, Tennessee. Note comparison to half-quart fruit jar.

Photo courtesy Museum of Appalachia, Norris, Tennessee

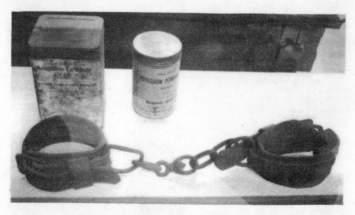

Set of animal hobbles, veterinarian's, with tools and medicine containers.
Photo courtesy Museum of Appalachia, Norris, Tennessee

in extended bundles using thin wood splints, these strands interwoven to connect as the basket developed. Coil size depended on the finished object, be it bread-dough riser, fruit basket, or beehive. Ryestraw containers seem to have been valued, and a number are preserved.

Wheatstraw, shorter and sturdier, was once made into doll-figures and small animals, some quite well done. Another pastime was working individual straws into geometric forms, sometimes as Christmas tree ornaments, but few such examples are left for study.

Values

Ryestraw wringer or press, all wood, two rollers in frame
    with crank handle, about 16 in. high $65-80
Ryestraw basket, 6 in. diameter, bowl-type $16-20
Ryestraw basket, bread-riser, 8 in. high, 14 in.
    diameter $35-45
Doll, wheatstraw, rough, early, gingham dress $9-12
Geometric figure, wheatstraw, traces of paint, 1¼x1¾
    inches, worn $4-5

Stray Preventers

When an animal got out, finding lost stock was only part of the problem. The farmer had to make amends to neighbors for damages done to their crops or property. The potential loss could be more direct, for cows and horses tended to over-eat and "founder"—become distended with too much strange food, with serious health consequences.

Beyond fencing, a number of devices were used to keep animals at home, or at least nearby. Hobbles were attached to the lower legs and kept the animal from running. Tethers were ties of many kinds, usually temporary. Knee-knockers were wooden weights suspended from the neck, which made mules and horses move at a slow pace.

More common, however, are the primitive "fence-pokes" or "cow-pokes", made for almost every cloven-hoofed farm animal. These were mostly of wood, neck-collars with projecting prongs that prevented the animal from thrusting through spaces in a rail fence. Some also kept the animals from jumping wire fencing, as the top strand of barbed-wire could cause serious injury.

Values

Hobbles, chain connected (13 in. long), heavy leather
    shackles $8-11
Knee-knocker, chain-suspended wood cylinder 16 in.
    long $20-30
Fence-poke, calf-size, primitive, six projections $25-35
Fence-poke, cow-size, two top/three bottom projec-
    tions $30-40

Tanning

Tanning was an early skill that many farmers had to know. Tannage relates to the lengthy process of turning fresh rawhide into leather, and several steps were involved. The word itself comes from tannic, the substance derived from certain tree barks and used in the soaking/curing process.

Bark-spuds or slicks were giant chisels used to remove bark, and the oak family provided most of it. This in turn was crushed in large horse-powered mills or small hand-powered grinders. The tan-bark was placed in pits or vats with water, and the hides were weighted and soaked in them, with regular stirring.

The actual worker, who had a slanted large-beam table, was called a currier. He used two curved-blade implements, one to remove excess tissue, another to dehair the hide. As the hide slowly cured and darkened, it became leather, to be worked and oiled to become supple.

Values

Bark spud, steel head 3¼ in. wide, 19 in. oak handle... $36-42

Bark mill, general farm grinder, hand-cranked $18-24

Currier's knife, wooden handles, incurvate blade, 14 in. wide.. $22-29

Currier's bench, peeled log with two pegged legs at higher end.................................... $35-45

Thermometers

Thermometers are heat-measurers, and farms had several outdoor types. The most common and obvious were advertising thermometers that the farmer located very carefully so sun or shade did not give a false reading.

Most were stamped, tinned metal with advertisements in raised lettering and contrasting colors. The thermometer was a glass tube set somewhere toward the center of this miniature billboard. Providing the farm with free thermometers, the advertiser sold its products and the farmer learned how cold it was in winter or how hot the summer had become.

The one temperature mark watched most carefully was the freezing point of water, the "frost-line", at 32-degrees F., 0 at C. This could mean damage to garden seedlings and orchard buds in the spring. In the fall, freezing was welcomed as certain heavy work like logging was made easier, especially with snow-fall.

Values

Cream-separator advertisement, wood backing, white paint, light flaking, 5x14¾ inches $23-28

Camel Cigarettes, raised pack, fine condition $30-35

Chesterfield Cigarettes, raised pack, some paint drops . $25-30

Hires Root Beer, bottle-shaped, 27 in. high $40-50

Pepsi-Cola, bottle-shaped, about 2 ft. high $40-50

Tool Carriers/Holders

Today in our service-oriented economy, a man may carry his "tools" or specialized skills and knowledge largely in his head. Once tools were more an extension of the hand and needed to be transported and stored. So, a number of farm containers relate to tool carrying and storage.

One very common farm carrier was tray-like, with a high central grip or bar, used to hold hammer and nails. The latter were often placed in individual wood-walled compartments, segregated as to size or purpose. Kept at a handy location, the carrier was needed for fence-fixing and "tacking back" loose boards on barn and outbuildings. Horse-shoeing kits were common, so also the various bags carried by veterinarians.

Tool storage was important both to protect the tools and prevent loss. Carpenter's chests for woodworking tools are still to be found, though stoneworking tool-chests are scarce.

A surprising number of early chests—and the original tools, if still in them—were made in England or Europe and brought here by tradesmen immigrants.

Values

Nail and hammer carrier, wood, five compartments,
    10x21 inches $15-19
Veterinarian's bag, leather, with several syringes of stain-
    less steel $11-15
Farrier's stand, with hammer and hoof-paring knife ... $40-50
Carpenter's chest, empty, well-made, 16x30x21 in.
    high $80-95
Stoneworker's chest, several early tools, lift-out top
    drawer with half-a-dozen punches $120-135

Trapping Supplies

Trapping fur-bearing animals was always a part of the rural scene, whether farmboys primarily sought extra cash or the farmer wanted to rid his land of potential predators. About the only animal trapped only because of its pelt was the muskrat, present in the uncounted millions.

Even then, it was considered a nuisance because it weakened the retaining walls and banks of farm ponds and lakes. Rabbits too were trapped, for they had a late-winter habit of nibbling the bark of young orchard fruit trees.

Other furbearers were trapped, either because the coat was worth money from the local fur-buyer or because they were thought to pose a threat to the domestic fowl in the chicken coop. Among animals known to take or kill chickens were: Raccoons, skunks, opossums, weasels, foxes and coyotes. Also targeted were occasional mink and fisher, plus the bobcat.

Traps used were metal, and early types were handforged in many styles and sizes. Each trap had a bait pan or pressure plate, which, when depressed, released a lever which allowed the spring(s) to force the jaws together. Varieties of small-animal traps have one or two leaf-springs, single or double jaws, and were usually set on the ground or in the water.

Others, like the jump-traps, had a spring below the trap to hurl it into the air for a better leg-hold, while the "breaker" trap had an extra lever that struck above the jaws to cause further injury. Some traps had a coiled, compressed spring, perhaps combined with screw-on serrated teeth for the jaws.

Values

Trap, "Newhouse", (Oneida, N.Y.), double-jaw, with chain, single-spring, rusted $8-12

Trap, "Victor/No. 1½", chain, single-spring $4-6

Trap, jump-type, "No. 0", chain, single-spring $5-7

Trap, bear, "Herter's/#6", very large size $300-350

Trap, bear, cast and wrought iron, "No. 15" on trip-pan, 35 in. long $145-175

Trap, beaver or wolf, hand-forged, marked pan....... $150-175

Trap, hand-forged, early, unmarked, large $65-80

Clamp, for setting spring, screw-type, "No. 5" $13-17

Clamp, spring-set, lever-type, 11½ in. long $9-12

Pelt frame, hide-drying, metal strips on edges, 18 in. long $12-16

Pelt board, raccoon-size, tackmarks, weathered $4-7

Pelt boards, matched set of three sizes, refinished $15-20

Lock and key, for old wrought iron bear trap, 7 in. high $50-60

Range of hand-forged and factory-made traps. At lower near-left is an early fur-buyer's broadside exhorting people to trap, headed "GET BUSY / Furs Up And Booming".

Photo courtesy Museum of Appalachia, Norris, Tennessee

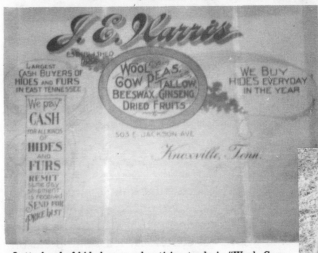

Letterhead of hide-buyer, advertising trade in "Wool, Gow Peas, Beeswax, Ginseng and Dried Fruits".

Photo courtesy Museum of Appalachia, Norris, Tennessee

Tool carryall, just under 3 ft. long; such types were often used by the farrier for carrying handled tools. This example is made of pine.

$45-55

225

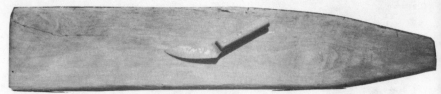

Skinning accoutrements for fur-bearing animals. Center, skinning knife 8 in. long. $7-10

Background, pelt-board, raccoon-size, 32 in. long. $10-14

Bark mill, horse-powered in circle around central beam; the crushed tanbark was used for processing leather.

Photo courtesy Museum of Appalachia, Norris, Tennessee

Travelers

Travelers were the small wheels that made large wheels possible, and they actually had two purposes. The typical traveler was a wood or metal circle up to a foot in diameter, with handle secured to the wheel center. At least one point on or near the wheel rim was marked, and a few were scaled in inches.

Turning freely on its axle, the traveler measured circular distance accurately. The first use was in determining sectional lengths when a wooden wheel was assembled, with the last fit the most important. The traveler was also used repeatedly thereafter to measure the correct fit of the iron rims or tires to be shrunk onto the wood.

While travelers were commonly associated with the professional wheelwright or wagonmaker, the occasional early farmer used one for wheel repairs and tire replacement.

Values

Traveler, wheelwright's; has pointer, wheel 8 in. diameter $33-38

Traveler, brass wheel and pointer, wood handle, 7⅛ in. diameter $50-60

Traveler, hand-forged, wheel 5½ in. diameter........ $26-32

Traveler, all-wood, solid wheel 8¾ in. diameter, hand-carved except for lathe-turned wheel $45-60

Treenware

"Treen" or treenware are objects of solid wood, usually hand-carved, often quite old. The name comes from the original wood source, the tree. While not all the items listed below had a direct farm use, all could indeed have been made on the early farm. Note that some of the pieces were lathe-turned, still considerd treen by most authorities.

Values

Bowl, exterior blue and bottom branded with initials; 7¾ in. diameter $75-85

Bowl, interior red, exterior green, 7¼ in. diameter $65-75

Bowl, interior with paint traces, 3¼x12¾ in. diameter $100-115

Box, turned foot and lid, 7½ in. high................ $60-70

Butter paddle, with age-crack in bowl, 9 in. long $45-50

Butter paddle, 7½ in. long $28-32

Canister, turned rings with brown stain, inset bottom, 9½ in. diameter $55-65

Chalice, turned, two-color, 5 in. high $25-30

Dipper, thin, well-shaped handle, tin repairs of old age crack, 4¼ in. diameter $195-225

Dipper, handle hook for edge of pot, 10¾ in. long $60-65

Goblet, footed, old red paint, carved, 3½ in. high $40-45

Ladle, all-wood, bowl with repaired crack, 13½ in. long $27.5-35

Yarn-holder, turned, rim-hole for yarn, 8¼ in. high ... $205-235

Scoop, treen, one piece with long handle, miniature,
5¼ in. long $80-100

Wagon Jacks

In horse-drawn days each farm had at least one wagon jack for maintenance and repair of wheeled vehicles. They were used to raise one axle-end to remove a wheel for rim, spoke or hub repairs.

Early wagon jacks tend to be all-wood, operated by lever action. Sometimes the mechanism had cogs or steps to control height, plus a safety wedge to prevent slippage, or a catch-lock to hold position.

Most early jacks had a "lift" of only a few inches, though some later all-metal manufactured jacks could raise several tons as much as 18 inches. This stronger jack type, while designed for heavy farm equipment, was also used by farmers for other tasks. They could help position heavy machinery, even lift the sides or corners of small buildings for leveling or foundation replacement.

A number of the early wagon jacks, of wood and wrought iron, were intricately made. Some are considered true primitives and are widely collected.

Values

Wagon jack, operated with iron crank-key in side, cogged wrought-iron wheel, 22½ in. high $40-50

Wagon jack, lever action, cast-iron, 25 in. high $32-40

Wagon jack, wood with wrought iron fittings, dated,
23 in. high $35-50

Wagon jack, iron screw type, tree-trunk base, 21 in.
high $35-45

Wagon jack, cast-iron, lever-type operation, maker-marked, 24 in. high $25-35

Wagon jack, very early, wrought top-catch and cogged interior wheel, crank action, 26¼ in. high $45-55

Walking Sticks

Though certainly not relegated only to farms, sticks used in walking were helpful adjuncts to farm life. Plainer versions, the canes, were used by the old or infirm, but the walking stick of about one yard in length had other purposes.

The walking stick could measure distance or height, was an aid to walking plowed ground and hillsides. It could kill a snake, be rapped on pond ice to determine thickness, be thrown to bring down nuts. Some sticks were more status symbols than practical friends for a walk around the farm. Many seem to have been owner-made, a form of genuine folk art.

Values

Walking stick, carved, with snake around shaft, 35½ in. long $25-40

Walking stick, eagle with serpent in beak, 36 in. long .. $55-70

Walking stick, branch, folk art with face on grip. Worn, 38 in. long $55-70

Walking stick, made from root, duck-head top, 34 in. long $45-60

Walking stick, twisted shank, pig's head handle, 36 in. long $75-95

Wash-Basins

Wash-basins or wash-bowls were the hired-man's friend, and the threshing crew's salvation. The shallow water containers were used to "clean up" men who had "come in", that is, arrived from the fields for a hearty dinner or to end the day's work.

While the wash-basin was usually hung near the kitchen door or on the out-kitchen for general face and hand washing, the threshing crew required an assembly-line setup. Necessary were a long board or low table, and two wash-basins. One held hot, soapy water, the other cold (often direct from the well or spring-house) rinsing water.

Ex-feedsack drying towels completed the ritual, and the famished men sat down to a meal that was usually a multi-splendored thing. And as often as not, one of the wash-basins had been used with a damp cloth cover to help "rise" the bread the night before.

Values

Wash-basin, navy blue and white speckled, part of
original label, "Enameled Steelware/Newark, N.J.",
holed to hang $20-25

Wash-basin, blue and white "spongeware", 11 in.
diameter $11-16

Wash-basin, cobalt blue and white "agateware", 9¾ in.
diameter $13-17

Wash-basin, early tin, rim-hole, 13 in. diameter $19-24

Water Carriers

Though seemingly mundane, water carriers were all-important in allowing farm workers to maintain efficiency. This is one area of farm-related collectibles in which easily transportable water containers were not always made for that task alone. Many of the containers had originally contained some other store-bought liquid.

Water carriers had some sort of handle, and were of wood, metal, glass or stoneware. Personal canteens of wood were sometimes carried, and bentwood field kegs can still be found. Stave-constructed examples were more common, with hole at center or in one end.

One gallon sizes were popular, and many a farm lad became the "water boy" for the duration of haying season. The glass jug that might once have held vinegar or cider became the ubiquitous farm water carrier, complete with a glass neck loop for additional finger support. The experienced farm worker lifted it with one hand, the weight resting in the fork of the upraised arm and shoulder.

The harvest ring was a very early and rare ceramic circle, and old illustrations show scythers with the ring around the neck and under the opposite arm. But, since the only thing worse than working in hot weather was also drinking warm water, this carrier probably was stored in the shade.

Values

Canteen, stave-constructed, wood-strapped, ca. 1850 .. $70-85

Water carrier, rectangular, tin, bail handle, 14 in.
high $16-22

Water jug, clear glass, half-gal. size, embossed label of whiskey distillery $18-25

Stoneware jug, cobalt-blue butterfly, one-gal. size, old corncob stopper $100-125

Water keg, redware with clear glaze, about 13 in. high, center bung-hole $165-190

Harvest ring, grey pottery, 15 in. exterior diameter $200-plus

Water Pumps

Long-handled suction-powered wood or metal pumps to lift water from drilled well or dug cistern were features on every farm. Some late 1800's pumps were mostly wood, even with hollow wood water-tubing. Though few of these have survived, they are highly collectible. Many cast-iron pumps could be operated either with the long lever-arm, or be windmill-powered; some could lift water from wells as much as 150 feet deep.

A variation, set over a dug well or cistern, had a wood or metal housing and was operated by a side crank handle. Rather than the usual up-and-down piston stroke, the water was pulled up with an endless chain of small buckets or rubber plugs in a tight-fitting pipe. Such "chain pumps" are sometimes offered for sale, especially the casing which may be of stencil-decorated wood.

The most likely pump to be seen is the much smaller "pitcher pump", most standing about 16 inches high. Designed to be mounted off the floor, these were favorites for the closer quarters of kitchen or milk-house.

Transportable water pumps were sometimes used, of cast-iron, with a direct-action plunger. Set up beside any water source, they could be used for washing windows and farm equipment.

Pump, pitcher; cast-iron, orig. red paint, complete $25-35

Pump, crank system with suction cups and chain, 38 in. high $65-80

Pump, crank or chain-lift, small rectangular buckets, most intact $75-90

Pump, carry; "D"-ring handle, 26 in. high, with one hose still attached............................. $18-23

War Effort

The advent of WW-II had a profound impact on the American farmer, and on farm-related items today. Farmers—those not caught up in the massive draft—were asked to work harder, produce more. They were requested to collect milkweed pods as a substitute for kapok in life jackets and save grease for certain munitions, and conserve. They did.

Farmers were also asked to collect and turn in scrap metal. Many old tools, from hand-wrought knives to complete pieces of equipment, went to the collecting centers at a few cents to the pound. Many of the metal objects really had outlived usefulness, or had been replaced by efficient devices that made the collected objects obsolete.

Still, one wonders, what elegant treasures were made into bomb-fins and artillery shells. What artful creations were reduced to molten metal to go abroad in new and deadly forms? The writer does not begrudge the necessary national effort, but from an historic point-of-view, there is no getting back what was lost in the period 1942-1944.

Weathervanes

Weathervanes are slightly misnamed, as they do not directly give present or future weather, only the current wind direction. The first weathervane symbols, originating in Europe, were in the form of roosters, and were known as "weathercocks". No large barn was truly complete without an impressive weathervane at top center.

On American farms, the rooster often gave way to other figural forms, such as horses, pigs, even maidens, Indians and sailing ships. Weathervanes are always set off-center, so that the wind will press more firmly on the back part of the figure. then it heads into the wind, showing the exact direction of the invisible gusts.

Mounted on a stationary rod or firm base, weathervanes were made of many materials—iron, cast and/or wrought, wood, tin, and other metal. Some of the more valuable types are hollow-bodied and copper-skinned, made up by skilled craftsmen and small firms in the late 1800's. As the advanced collector well knows, a number of very impressive reproductions exist, sometimes offered as both old and authentic.

Windvanes or "directionals" are similar to weathervanes only in that they swing easily in the wind. Inexpensive and often of cast-iron, they typically have a broad wind-catcher in the shape of an arrow, plus indicators for the four directions. Some forms even incorporate a lightning rod.

Values

Weathervane, tin, primitive, directional and finial, 24 in. high	$95-125
Weathervane figure, wood, of running horse, old black paint, 25¼ in. long	$235-255
Weathervane, sheet iron rooster figure, 22 in. long	$150-175
Weathervane, iron, zinc running horse, 33 in. long	$105-125
Weathervane figure, copper, eagle with good patina, wood base, 24 in. wingspan	$650-750
Weathervane, copper and brass, arrow and ball, 23½ in. long	$90-110
Directionals, milk glass ball, zinc and cast iron, 15 in. high	$45-55
Directionals, cast iron, arrowhead tip and finned wind-catcher, 17 in. long	$15-20
Directionals, wrought-iron, scroll with brass letters, 85 in. high	$175-200
Weathervane arrow, cast-iron, cobalt blue glass in tail, 18½ in. long	$30-40

Wheels

Farm wheels ran the gamut from oldest all-wood to mass-produced examples with metal-lined ("ironed") hubs and iron-strapped rims. Then came the all-iron tractor and implement wheels, some with projecting lugs which tore up county roads. Finally, the rubber-rimmed wheels came into favor. Of them all, wheels with wooden hubs, spokes and rims are the easiest to obtain, the most sought-after.

The largest classes are the heavyweight wagon wheels, and the lightweight buggy wheels, both marvels of specialized skills. They combined shaped and seasoned hardwoods with the ability to apply wrought-iron in needed places. This was the special domain of the wheelwright. The writer was priviledged to sort through a family-owned farmstead purchased from an early 1900's wheelwright and farm wagon builder.

The three outbuildings—chicken-house, corncrib and wagon shed—were literally packed to the rafter corners with goods. These ranged from general farm items to wheel-making tools, and there were perhaps three dozen completed wheels and hundreds of wheel parts.

In addition to several matched pairs of wagon-wheels later sold to a clothing store owner for tie-display racks, a number of interesting tools were found. There were several draw-knives, a large device (wrought-iron, lever-action) for setting spokes in the hub and rim, a "traveler" for measuring the outside wheel rim, a spokeshave, a spoke-pointer that resembled a large hand-twist pencil sharpener, and several unknown tools apparently designed for shaping or thinning wood strips.

Values

Wagon-wheel, complete, 31 in. diameter	$20-25
Wagon wheels, matched pair, 29 in. diameter, the set	$45-55
Buggy wheels, matched pair, 42 in. diameter, the set	$55-65
Spoke-spacer, wrought iron, 23 in. long	$75-95
Traveler, wooden handle, wrought wheel, 12½ in. in diameter	$45-55
Spoke-pointer, black japanned metal, 8 in. long	$28-32
Spoke-shave, drawknife with small curved blade, cast-iron	$23-28

Whimsys

Whimsys are quaint and fanciful works, often handcarved of wood. While some were obviously created at the spur of the moment, others evidence both care and planning and are quite intricate. Often it would seem that the maker wanted both to show a command of tools and yet make something attractive, if not useful.

So there are whimsys as touching as a love token, and as useless and decorative as endless-link wooden chains. At one time, peach pits were even carved into miniature baskets and faces, to be used as part of something else or complete miniatures in themselves.

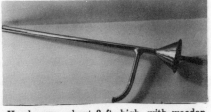

Below
Swift, adjustable wool yarn winder and for measuring; note base-clamp for securing to edge. $110-140

Hand-pump, about 3 ft. high, with wooden plunger-lifter. Water was pulled from bottom hole and emerged from curved pipe at top. All-brass, polished and lacquered.
$90-110

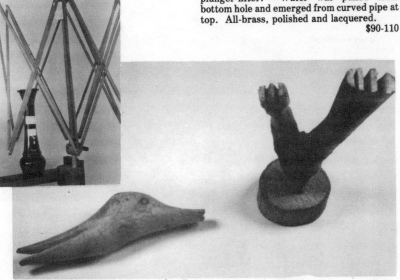

Whittler's art; left, bird's head about 4 in. long. $6-7

Right, rooster made from forked limb, on section of cut sapling; by Russell Duncan, about 3½ in. high. $7-9

Yarn-winder, for wool thread, block base, about 32 in. high.
$85-105

Grain sieve, bentwood frame, about 15 in. diameter. Note the "pie section" form of three wire braces.
$12-15

Grain sieve, bentwood frame, wire mesh openings approx. quarter-inch, about 17 in. diameter. $13-16

Whimsey creations, whittled. Left, cat figure, 3 in. high including treetrunk base. $5-7

Right, mushroom group made of butternut tops, whittled stems and trunk base. Similar examples are still made today. $6-8

Rooster weathervane on cupola, with wind-arrow and directionals.

Farm pump, cast-iron, about 36 in. high; marked "HAST - FOOS & Co. / #191". $17-22

Grain-threshing maul, all wood, primitive, from Minnesota, 7 ft. handle. $50-60

Photo courtesy Fairfield Antiques, Lancaster, Ohio

Values

Whimsy, carved wood, caged spheres, 13½ in. high ... $105-130

Whimsy, turned wood, box with lid, 7⅝ in. high $15-20

Whimsy, carved chain with large wooden padlock, all
    from one piece of hardwood; age patina $80-95

Whirligigs

    Whirligigs or whirleygigs have become top collector pieces in
the past few years. The typical example was well-made of wood
scraps, and pole-mounted to catch the wind. Figures were
activated by windmill-like fans or oar-like vanes that—
connected by an intricate drive system—made cartoon-like
figures move in endless repetition.

    The motif might be patriotic (Uncle Sam saluting), historical
(Indian paddling canoe), everyday (character(s) sawing log), or
humorous (farmer milking kicking cow). Less detailed versions
might be made on the farm itself, though finer examples were
probably turned out by a local craftsman.

    Whatever the source, whirligigs offered innocent amuse-
ment, stationary fun activated by the wind. Also known from
early times as the "spinning coach", whirligigs may have been
in the form of a horse-drawn enclosed carriage with turning
wheels.

Values

Whirligig, 20th century folk art, Uncle Sam, revolving
    vanes on ends of arms, small repair, 15½ in. high,
    plus base $145-170

Whirligig, man milking cow, contemporary, polychrome
    plywood, hind leg broken out of nail, 18¼ in.
    high $45-60

Whirligig, wooden, sailor with semaphore flags,
    weathered blue and white paint; 20th century, 9¼
    in. high $100-125

Windmill Weights

    (The writer wishes to thank Mr. Edwin Wingfield, "The
Windmill Man", Hamilton, Illinois, for the following information contained in a personal communication.)

"I have run across many (of the windmill weights) in the 28 years I have been working on windmills.

"Windmill weights were used to control the speed of windmill wheels in strong or high winds. They were so engineered that the stronger the wind blew, the more the weight would have to be lifted. These weights were on the end of a pipe about two-and-a-half feet long.

"The weights themselves were mostly in the form of heavy solid balls about three inches in diameter. Also some were in the form of animals, generally horses.

"I still see some in the area in which I live in Illinois, one mile from Iowa and two miles from Missouri. I have always heard the weights made of heavy cast metal, and in the form of animals, were made in Germany."

Values

Windmill weight, cast rooster, painted white; has
    remains of wooden base, 9½ in. high............ $135-150
Windmill weight, cast-iron horse, 16¼ in. high $185-200
Windmill weight, cast-iron horse, 16½ in. high $240-265
Windmill weight, ball type, 2 ft. high $100-125

Winnowing

To winnow—a word perhaps developed from "wind-throw" —was to separate grain from chaff using wind or air currents. Entirely hand-done, the process began when ripe heads of grain were pounded to release the grains.

A common pioneer device, the flail, was simply a medium-length pole with a short club-like piece secured by a leather thong. Heavier poles without attachments were also used, the aim being to hull the grain without bruising it.

Winnowing, on a breezy day, entailed standing upwind and throwing the chaff-grain mixture straight up into the air. The heavier grain fell back, the light chaff was blown away. Flat, wide scoop-shovels might be employed, though the favorite device was the winnowing tray.

This was a two-handled broad scoop with wide lip and bentwood sides and back, 2 feet and more across. Relatively fragile, they are rare today. Winnowing screens and baskets were also used, and all are wide and flat.

Values

Flail, hickory handle, later wire-connected club, handle
7 ft. long $25-35

Flail, very early, 6 ft. handle, two slats attached to far
end. $30-40

Winnowing club, 5½ ft. long, limber for striking $15-19

Winnowing tray, slat bottom, slight damage to lip, one
handle missing $40-50

Winnowing basket, circular, bentwood frame 3¾ in.
high, 25 in. diameter $45-55

Wool
(See also Linsey-Woolsey)

Sheep, as soon as the farmer could afford to fence his property, became important livestock, both for meat and wool. Sheep could nip down sparse weeds, eating what cattle and horses did not. Lambchops and mutton brought good prices from the meat-buyer and butcher, and lambskin and sheepskin could be made into leather or warm garments. Even Western ranchers raising cattle swore by their sheepskin coats in wintertime.

Working with wool bundles after spring shearing had one good and one bad side. Lanolin or "wool-fat" was a great help to sore and cracked hands, but the ticks found in the wool were likely to attach to the nearest warmth. Wool for sale was first gross-weighted, then graded by "staple", the strands judged on length and fineness. Wool that was thick and long brought the best prices.

While some meat and leather was consumed on the farm, an interesting and still-valuable commodity resulted from the interaction of raw material and manufacturing innovation. Several businesses processed bulk wool into all-wool rugs, receiving and sending by rail freight. Some of these fine rugs survive in good shape, even after 75 years of rigorous use.

A number of wool-related collectibles exist, some indirectly a product of sheep-grown fiber.

Values

Sheep-shears or "clippers", spring-handled steel, 13 in.
long $5-8

Clippers, crank-operated, jointed revolving extensions,
hand-powered, two-man, ca. 1905 .... $20-30

Niddy-noddy, for winding wool yarn, hand-pegged
maple, wrapping heads at right angles, single
shaft.. $53-65

Spinning wheel, for wool, wheel 49½ in. diameter $135-150

Spinning wheel "finger" (thread distributor), of lignum
vitae wood, unusual.......................... $16-20

Yarn-winder, floor type, turned legs, chip-carved, 40 in.
high $75-95

Yarn-winder, four tall legs, four arms, pegged, 37 in.
high $125-150

Wool-carder, wood, worn iron teeth, pair, for pulling
and cleaning wool strands for spinning and
weaving..................................... $25-30

Rug loom, table-top size, 4 ft. wide, all wood $60-75

Basket, for washing wool, four pegged feet, handled,
splint construction, 17x24x15½ in. high $90-115

Rug, floral border, subdued reds and blacks, good pile,
farm-raised wool, size 9½x12½ feet $115-140

Wrenches

Devices for grasping and holding or turning other objects, wrenches in many sizes could be found on all farms. No matter the specific type, all were handled, for leverage, and the jaws operated in one of two ways. Either they were set to a certain size for one fit, or they were adjustable to perform many tasks.

Strongly constructed pipe wrenches could be used for other, related work, as could the lighter expanded-jaw "monkey" wrench for turning metal nuts. Other popular wrench types include the ratchet, socket (often in sized sets) and tappet wrench. Metal wrenches were especially useful for keeping all farm mechanical equipment in operating condition.

Values

Wrench, adjustable, "Coes Wrench Co.", 7 in. long ... $12-14

Wrench, adjustable, "Keen Kutter/Pat. 1896", 22 in. long $23-27

Wrench, adjustable, "Girard Wrench Co.", 11 in. long $13-15

Wrench, double-head, "Case Eagle", 20 in. long $24-27

Wrench, monkey, 3 in. jaws, handle 11 in. long, un-marked $4-7

Wrench, S-shape, double-end, Winchester-marked, 10 in. long $20-26

Wrench, chain, "Stansbery/Pat. 1901", 29 in. long $34-39

Farm Outbuildings—A Description

When Eastern opportunity seekers moved Westward into the hardwood forests and canebrakes and grassy glades, they did not first build log cabins. According to a number of chronicles from the mid- and late-1700's, a single or married man made the long and dangerous traverse alone or in company with other men. They prepared the way. And, in the first examples, farm animals did not normally accompany the men.

A log cabin—often the beginning of a farm—took weeks and months to put up and complete. Hopefully sited near a spring, the cabin required trees to be cut down, seasoned somewhat, cut into lengths, broad-axed into squared logs, and notched at the ends.

When finally "laid up", the horizontal beams had to be "chinked" with a daub of mud and vegetation or horsehair to close off drafts. Roofing, chimney, fireplace, foundation, doors and windows were other matters.

Until all this could be done, the builders had to live somewhere. The first permanent settlers—taking title by land-grant or outright purchase—lived in somewhat basic shelters. One was the lean-to, borrowed from the experience of Native Americans. It was often a pole-and-bark affair, erected in a hollow or against a rock outcropping.

Another pre-cabin shelter was the sod-covered bunker, perhaps constructed partly underground. Eventually the workers moved into the then-luxurious log cabin, complete with all the hearth-warming things one could make. Any animals were also protected by makeshift shelters before a log barn was constructed.

Next, the land had to be developed and exploited. While the cabin changed to become clapboard-sided frame dwellings, and imposing edifices of brick, stone or wood, the early farm required more. Outbuildings became necessary, from barn to springhouse to wagonshed.

The first out-buildings—originally, any structure except the home, later, anything except the house and barn—were used to shelter animals. The more important farm animals were oxen, mules, horses and cattle. Eventually, the barn developed, to become an extraordinarily large structure designed for multi-purpose use. Whatever the regional style or size, it was put up

242

to protect the farmer's investment in livestock and the feedstuffs needed to keep them.

A sprinkling of barns decorates the rural American landscape. Whole books have been done on them, architects have blessed the functional purpose, painters have specialized in their portraits. Though a good dictionary states that the word "barn" relates to the keeping of "barley", it is the main storage unit for farm crops.

Several dozen different barn styles developed, and the layout varied depending on the use. Barns were built in the open and above-ground, but many were set into a slope, for several reasons. The earth gave the first level warmth, while the upslope level made it easier for horses to pull loaded vehicles into the second storey.

This was the "wagon floor", located over the livestock pens and stalls. Most barns had several levels, plus other high haylofts or "bays". In outline, most barns were rectangular, but square and round examples exist. Other barns had from five to eight sides.

A typical barn contained numerous sub-stations beyond the sections sheltering animals. There were large feed-bins for cattle, and horse-stalls with long manger and high, side feed-boxes for corn or oats. If milk-cows were raised, the milking area attended to their needs.

Tack-rooms were sometimes included for harness and accoutrements; a number of farms did not have a separate out-building granary, and included one beside the second-level wagon floor. This was usually in the form of a long, narrow room with the barn wall along one side and a hayloft overhead.

The other side and far end contained doors to large bins with special openings. These were closed by loose boards, which fitted into frame slots at each end. The number used depended on the height of the loose grain in individual bins.

Corn was rarely stored in such in-barn granaries, and the "small grains" were binned. Oats, rye, barley and wheat were put away, and the average number of these large bins might be four or five. Such storage kept the grain with low moisture-content, and it was readily accessible for animal feed and mixes.

About a hundred years ago, a new feature was added to some barns, this being the feed cooker. Since raw grains were valuable and not suited for all young farm animals unprocessed, large kettle-like cookers began to be used.

Often steam-heated and usually set up on the ground floor, nourishing "scientific" mashes could be cooked. Sheep were sometimes kept on the ground floor, even chickens in a part of the second storey. But pigs had their own special areas and building, and the old-fashioned pig-sty became the hog-house.

Special-purpose barns were also put up by farmers having practical knowledge in some area. So, there are barns designed for storing and curing tobacco, for dairy purposes, for horses (stud farms), for sheep, for meat animals. An aware person, walking through almost any barn, can know a great deal about what the farmer was thinking and planning when the place was put up.

Beyond the barn, many other out-buildings were erected, each to serve an economic function at the time. As times changed, some buildings became obsolete and were torn down or fell into uselessness and disrepair, or served other purposes. And with them, sadly, went vast portions of a whole way of life.

Smokehouses or "meat-houses", aside from the privy and one or two other out-buildings, were the smallest. They were tall, since smoke rose and meat had to be hung high in the fumes to cure and take on flavor.

There was always a provision for coals, and these and smouldering hickory or corncobs could sometimes be removed via an outside opening. The smokehouse, whether frame with wood-shingle roof or of stone or brick, followed-up the butchering process, preserving the meat in a special way. Occasionally a smokehouse was put on the second floor of a milkhouse or springhouse.

Drying sheds, not found on most farms, were unique structures. Designed to slowly heat and remove water from fruit and vegetables, these were small buildings with wide, flat dehydrating trays. On a warm, sunny day, the trays could be pulled out or spread on the ground, but drying in inclement weather meant that a fire might be put in. As the smokehouse preserved meat, so did the drying shed "put up" its own food for the future.

Springhouses have long been associated with farms, but they were formerly far less than a house-shape, and were not always set over a spring. While earlier farmsteads were indeed sited near year-round water-sources, the spring was only slightly improved, as the main demand was for pure water. The true springhouse developed later to also provide cooling services.

As the farm went from subsistance to production with surpluses, the refrigeration requirement came in. Shed-covered springs, to keep the water clean, were no longer enough. Constructed near the house or barn, the springhouse combined the old-fashioned ice-box and the present-day refrigerator.

Most springhouses, however they were constructed, had a bottom level with one deep part to provide bucket-dipped water. The main element was a long, shallow portion that cooled at about 55 degrees and could handle a dozen or more crocks. These in turn held dairy products like milk, cream, cottage-cheese and butter.

Also "kept" were cider and wine, and anything else the farm-family wanted to preserve, from leftovers to fresh-caught fish. It is interesting that the presence of a large and healthy crayfish in the springhouse water assured users that the water itself was good to drink.

The wagonshed was an omni-purpose out-building, relating to the farm and fields as did the garden shed to the family's personal growing plot. It might serve double duty as protection for surrey or buggy, and it was more a space worked into other buildings than a concept unto itself. If a farmer had many conveyances, of course, a separate building did go up. Simply, a wagonshed was where such conveyances were stored when not in use.

The wagon-floor of the barn was reserved for loaded wagons, for off-loading or storm protection. The wagonshed sometimes had ceiling or rafter-mounted rollers or pulleys for taking off the body so the chassis could be used for other purposes.

For granary purposes, especially on larger farms, a special farm building was erected to hold the "small grains". A typical granary had a central front door and a center-aisle that terminated either at a bin or a back door. Windows were few and small, and up to half a dozen large bins were compartmented into the structure.

In later times, pre-built metal containers were brought onto the farm for grain storage, including shelled corn. Until that time, corn was stored, unshelled, in the unique structure called the corn-crib. While often a separate long, narrow and high building, some farms combined twin corncribs, perhaps 18 or 20 feet apart, and covered the whole with a wide, center-peaked roof. The interior space was left for a wagonshed or other farm-equipment storage.

Corncribs can be recognized by the unique silhouette, and the fact that the exterior consists of narrow boards spaced an inch or moreso apart. This was done to facilitate the main purpose of the building, which was both to store and dry the corn. This allowed good air circulation, and some farmers even had corncribs with bottom-center tunnels which ran the length of the crib for more air-exposure.

Sometimes hardware cloth (metal)—or even wire-and-slat snow fencing—was run around the interior to keep more corn in and birds and rodents out. The corncrib was set up off the ground perhaps 18 inches, and metal shields were sometimes used to keep out rats. For whatever reasons, mice seemed to be the problem in granaries, rats in the corncribs and chicken-houses, while both coexisted well in the barn.

Hoghouses were generally set off by themselves, partly due to an odor problem, though space was sometimes made in the warmer barn for brooding sows and piglets.

Depending on how much the farmer "got into" chicken raising, but stayed with general farming short of a poultry operation, three different buildings might still be required. One was the "brooder" house, where chicks were hatched and/or raised. Some farmers put incubators in the farmhouse cellar, near the furnace, so the chicks could be watched more carefully.

Another chicken-house might have been the pullet-house, for fattening market meat or for raising hens. Chickens, pound-for-pound, usually cost less to raise than any other farm living thing. Some farmers made a sideline of raising ducks or turkeys, and other facilities were required for them. As a side note on the importance of raising and marketing chickens, a special poultry carrier could be purchased, that fit on a car's running board.

The tool shed was generally a small building or part of a

larger one. Later, some became the machine-shop, but at first the tool shed was where farm tools were stored, oiled, sharpened and repaired.

Along the same lines, some farms had equipment sheds elsewhere on the farm, often well away from the close-in gathering of out-buildings. Such structures, often near the fields where the appropriate equipment was used, were sometimes open on one side for easy access to horse or tractor.

If a large number of horses were used on a big farm, a special horse-barn might be put up to house a farmer's horse-power. Of course, two types of horses were always used on farms. In pioneer days, the riding horse might serve triple-duty, being also used to pull conveyances and farming equipment.

Later, certain lines or breeds were developed, and the lighter, faster horse were still used primarily for transportation. Larger, more powerful horses did the heavy work, and mules were often valued here. All were large, docile beasts capable of enormous output with reasonably good care. These were the "work" or "draft" horse for pulling heavy loads.

The forge-shed was sometimes part of the tool shed, but was always set off from the other large and flammable farm outbuildings. A scaled-down blacksmith shop, it was necessary for making and repairing iron and steel tools and parts.

The milkhouse could be a number of things, but always there was a workable system for cooling milk. Generally, it was an addition to the barn, typically a small block structure with cold, running water. Sometimes it was set off by itself, over a well or spring, or was placed beneath the windmill where the milk benefitted from the cold pumped water.

Almost forgotten today were two structures that once helped keep the hearth-fires burning. One was the woodshed, a frame structure that helped dry and season the many cords of wood once required. Protecting the cut wood from the elements, the woodshed usually contained a large chopping block for splitting wood. In turn, the wood might have come from land-clearing, or from the farmer's private stock of wood from the woodlot.

The woodlot was a nearby stand of first or second-growth timber, usually heavy in high-heat hardwoods like maple and hickory, and the oaks. Several acres in size, it was purposefully left uncut as a nearby fuel source.

After the shed that covered wood, coal came into popularity because it was relatively cheap, high in BTU's, and someone else had done most of the work in getting it. Some coal was piled, or put in the basement in a coal bin, but the well-run farm put the coal, like everything else, in a separate nearby out-building. Farm children soon learned that in order to be warm, one worked—either by bringing in coal or wood, tending the fires, or by carrying out ashes.

The ice-house deserves some comments. Preserving ice was as simple as flooding a low area of land in winter, putting in ice blocks and covering the heap with straw or sawdust, and earth.

The icehouse could be as elaborate as a Victorian icehouse with fancy gables and eaves. More likely—especially since it was set off by itself—it was a plain and effective structure that included dead-air-space walls, keeping heat out and cold in. It was not improved upon until commercial ice plants developed, along with cheap home delivery.

The out-kitchen or summer-kitchen still survives in some numbers. It was located very near the back door of the farm-house, and was generally put up about the same time. There is an error in wide circulation that a separate kitchen was required because farm families grew so large. While this may sometimes have been so, the summer-kitchen, by name, explains itself.

It was an outside kitchen used in pre-airconditioning days so that additional heat was not added to the house. Too, for harvesting or communal gatherings, large amounts of food could quickly be prepared, using both the house kitchen and the out-kitchen.

The structure, often with an over-hanging roof in front, could also be used for always-needed storage space, and some had an upstairs loft as sleeping quarters for hired hands. Many were used for making sauerkraut, to pickle cucumbers, or for homemade wine.

The pump-house, if there was one, was located beneath the windmill. It might include a trough for milk-cooling, or a spigot for family water; some had large holding tanks for water above them. After windmills went out of favor, the pumping station might include an electric motor that pumped water from the same source, a drilled or brick-lined deep well.

The sugar-shack, the distant structure that protected maple syrup and sugar-making operations, is almost unknown today. On farms that depended on trees for family sweetening, it was a minor building. If the farmer decided to make a commercial output, the sugar-shack might contain all the existing containers and tools, plus boil-down vats. Each sugar-shack had its own wood shed, or place where seasoned wood was stored.

Bake-ovens existed on large farms with large families. They were located close to the farmhouse and did the mass-baking not possible in the house kitchen or out-kitchen. Generally a building with walls and roof, the largest portion at one end consisted of the oven and chimney. Most disappeared with the advent of smaller families and better stoves.

The scale-house with a drive-on foundation became a feature on successful farms by 1875. Often it was located near the road so that others could use it for a fee. Some had pits beneath them for the scale mechanism, while others did not require one.

Sheep were kept for two major and one lesser reason. They provided wool and sheepskin for clothing, and mutton and lamb for the table or for sale. And, since they nibbled grass very close, sheep were ideal for giving lawns, pastures, orchards and meadows that just-mown look. Sheep were sometimes given a portion of the barn, especially for cold-weather and late-winter lambing use, but often had the ground floor of another building for living space.

The sheepshed or pen was mainly for nighttime use, and sheep were often allowed to forage during the day. The greatest danger to the flock was from sheep-killing dogs. This could be such a problem that special spiked collars were sometimes put around sheep necks, so that an attacking dog would injure itself. Killing mainly for sport, dogs could decimate a flock in a single night.

Most farms had an underground room known by various names, ground-cellar, cave-cellar or root-cellar. Typically built into the side of a hill, it could also be constructed on flat terrain by heaping up a mound of earth. A cave-cellar was an early form of earth-sheltered space, designed to keep certain foods cool in summer and prevent freezing in winter.

With floor level below the frost-line, a great variety of fruits, vegetables and root-crops could be stored for long periods with little or no deterioration. The rear of the below-ground room had a ventilation pipe to keep humidity down, and there were usually several steps downward inside the slanting front door. Long after more modern methods of temperature moderation developed, some cave-cellars were still used as storage areas.

A number of farms had a root-cellar constructed as the bank barn went up, and it was located beneath the wagon ramp to the second storey. The cellar typically was used to store root crops for animal feed, with a filler opening at the top, and a lower entrance to the room at the side or within the barn. In later years, many of these root cellars served as farm garbage dumps packed with now-interesting items.

Carriage-houses were related to wagon-sheds, and were popular on wealthier farms by the late 1800's. The structure related in a way to the wagon-shed, though the latter held the farm wagons and the carriage-house contained the family conveyance. Some buildings were quite ornate, especially in Victorian times. Eventually, the carriage-house evolved into the garage, just as the horse-barn or mule-shed developed into the tractor-shed.

A number of larger farms had ponds or small lakes, for fishing, skating and ice-cutting purposes. A popular practice on long-ago farms was a family picnic and row-boat outing, providing inexpensive nearby fun. The boats themselves were stored in pond-side boat house.

Now rare to the vanishing point, boathouses were simple frame affairs, with a wide, high door facing the water. Rowboats were generally stored on beams projecting from the boathouse sides, and turned sideways or upside-down to shed moisture. Traces of boathouse foundations can still be seen beside early farm bodies of standing water.

Interestingly, the first important structure put up on pioneer farms, the log or stone cabin, itself came to become an out-building. Very solidly built, and if near enough to the later farmhouse, some were turned into summer-kitchens or used for general storage space. A number, however, came to house the large family weaving loom, and were known as the loom-room or the loom-house.

Other structures, depending on different parts of the country, might be well-constructed lime-kilns or lime-ovens for reducing limestone to lime. This in turn was used for everything from fertilizer to serving as disinfectant for the privy to making whitewash.

Charcoal-kilns might exist, so also large drying sheds for orchards, and a partially-underground orchard-shed.

This is a look, then, at the early farms, the many-facted barn, the dozen and more specialized out-buildings. Each was designed for certain purposes, and they served those purposes well.

FARM-RELATED—General Listings

A number of values listings arrived after individual book subsections were completed, and, in some cases, there were not enough to warrant an individual class. Such examples are here collected together in order to make this book as up-to-date as possible, and include listings showing the wide range of farm items available. This section has been alphabetized for your convenience in looking up certain items of interest.

Values

Advertising fork and spoon, wood, each marked "Blue
Ribbon Dairy", the matched pair $11-14

Advertising match safe, "Sharples Cream Separator
Works" $35-40

Advertising medallion, brass, "International Harvester
Co.", half-dollar size $6.5-9

Advertising tip tray, "Dowagiac Grain Drills", grain
on wood background $20-25

Adz, iron, wood handle, 27 in. long $22.5-30

Animal collar, brass and leather, horse-like, 11 in. high . $5-8

Animal treadmill churn, 7 ft. long, working condition,
with wooden churn cylinder, staved $260-295

Anvil, for sheet metal, 18 in. long and graceful form,
cast-iron $105-125

Anvil, full-size, blacksmith's $65-85

Anvil, miniature, iron 3¾ in. long $18-25

Apple box, wood, orig. brown graining on yellow
ground, 9 in. square, 2½ in. high $95-115

Axe, mortise or post-hole, long thin blade $42.5-50

Bark spud, "slick", turned wooden handle, 29¾ in. long $35-40

Barking tool, chisel-edged, "Bangor/Maine", without handle $17.5-23

Barn heater-lamp, kerosene, tin, 10¾ in. high $35-42

Barrel, for cider jelly, small size $24.5-30

Barrel, staved, for water, bail single handle, 10 in. diameter $32-37

Basket, twin-based, very minor damage, 9x14x16 inches, plus splint handle $90-110

Basket, ryestraw, rim handle, 13 in. diameter $42-50

Basket, ryestraw, 11½ in. diameter $40-49

Basket, ryestraw, minor wear, 5½x11¾ inches $30-40

Basket, ryestraw, 4x11 inches $35-40

Basket refrigerator, splint with tin lining, ice compartment, 14½ in. high $25-35

Basket, splint, woven, double picnic-type handles, 11x15x28¾ inches $17.5-25

Basin, sheet brass, polished, 18 in. diameter $40-50

Beading tool, brass stop, good blade, 7½ in. wide $24-30

Bee smoker, copper, polished, wooden and canvas bellows, 9¼ in. long $85-100

Beetle, iron-bound wooden maulhead, orig. handle, 37 in. long $12-16

Bench, bootjack legs, old worn and weathered paint, 11¾x70½x18 in. high $125-150

Bit, bullnose; "Pat. 1887", 8½ in. long $18-23

Blacksmith's tongs, wrought iron, 17 in. long $9-12

Bob-sled, two sets of iron runners, from Maine maker, 9 ft. 3 in. long $45-60

Bolt headers, hand-forged, used to form bolt heads in blacksmith shop $20-24

Book, "Pennsylvania Folk Art", (1948), by Stoudt, John J. $60-75

Bootjack, cast-iron, pistol-shaped "The American bulldog boot jack", 8⅜ in. long $45-60

Bootjack, cast-iron cricket shape, 10½ in. long $20-25

Boring tool, "Grand Rapids Sash Pulley", 8½ in. long . . $23-27

Bottle, milk; "Farmer's Dairy/Mishawaka, Indiana" . . . $3-4

Box, apple; turned wood, 4¼ in. high $16-20

Box, butter; orig. blue paint, bentwood, 14½ in. diameter $35-40

Breaking bar, slater's; straightedge, for sizing slates, 18 in. long $23-28

Bucket, stave-constructed, 6¼ in. high $50-60

Bucket, stave-constructed, orig. yellow graining, 6⅜ in. high $60-75

Bucket, sheet copper, iron handle, 9¼ in. high $50-60

Bucket, spun brass, iron handle, 16½ in. high........ $55-65

Bucksaw, wooden frame, 16 in. long $27.5-35

Buggy whip, whalebone with ivory fittings, 72 in. long . $575-650

Bung auger, orig. wood handle, metal and wood, 6 in. long $17-22

Bung borer, unusual, with auger and four cutting blades, 16 in. long $45-55

Burl bowl, ash with excellent tight grain, small plugged hole in bottom and repaired rim crack; 6¼x19 in. diameter $325-400

Butter churn, round, platform base, crank handle, all wood $37-45

Calf weaner, nose-grips with protruding spikes, cast-iron 5½ in. wide $11-14

Calf yoke, bentwood, with pine crossbar $8-10

Cane, whalebone shaft with baleen rings and rosewood section, mid-1800's, 33 in. long................. $245-300

Caning vise, chair-maker's, maple, hand-pegged, wrought-iron clamps, 24 in. long $22.5-28

Carpenter's chest, poplar, interior liftout compartments, with wooden level; 19x20½x38½ inches $245-275

Carving, bearded man wearing turban, 5 in. high $25-30

Carving, folk art, wood relief, bird with real applied branches and carved applied leaves, orig. polychrome paint, 12x8½ inches $55-70

Carved man, horse-hair coiffure, 20th century, 13 in. high $62.5-75

Caulking mallet, with six caulking irons, graduated sizes $97.5-115

Chamfer knife, cooper's, "White/Buffalo", 19 in. long .. $45-55

Chamfer knife, cooper's, orig. handle, 16 in. long $40-50

Cheese corer/tester, steel semicurcular blade 14 in. long, with iron twist handle $24-28

Cheese shredder, hopper on board, crank handle, 10x11½ inches $52.5-65

Cheese slicer, on round wood platform, base-secured cleaver, side gear to push cheese wheel under knife, 20 in. base $200-250

Cheese tester, metal, "Moser Signau Rosterei/Switzerland", 6¾ in. long $16-20

Chisel, turning; gauged, delicate curved blade, 19 in. long $65-75

Chisel grinder, cylindrical stone, screw-adjusted angle, all metal $14-18

Chopping knife, steel blade, wood handle, fine condition $15-20

Churn, tin, complete with tin dasher, 13½ in. high $75-90

Clapboard gauge, metal, wood handle, and marked "Stanley" $16-20

Cleaver, butchering; 15½ in. long $35-40

Hatchel (flax comb), wrought iron nails, 6¾x16¼ inches $65-80

Cottage-cheese sieves, pair; tin, heart-shaped, 3¾ in. high, both $80-95

Cranberry picker, pine, iron "fingers", no finish, not too old, 16½ in. long $17.5-23

Cranberry picker, hickory handle, iron teeth, 43½ in. long $30-35

Cranberry scoop, early, tin and wooden prongs, 23 in. long $42.50

Cream separator sign, advertising, tin, with two attached small Jersey cows; sign 18x24 inches $100-125

Cream separator, on legs, cover with brass mesh screen, 39 in. high $55-70

Crock, stoneware, 2-gal. size, stenciled label in cobalt .. $150-175

Crock, stoneware, 6 gal. size, horsehead in cobalt, hairline crack, piece 13½ in. high $65-85

Depth stop, "Stanley/Pat. Oct. 11, 04", 6 in. high $11-13

Dipper, tin, bowl with wire-turned rim, hanging loop, 12½ in. long $15-20

Dipper, wrought iron, scrolled hanger, 20 in. long $30-40

Doorstop, cast-iron, elephant, old grey paint, 6 in. high $80-90

Doorstop, cast-iron, little girl holding skirts, "B & H", 11⅜ in. high $105-125

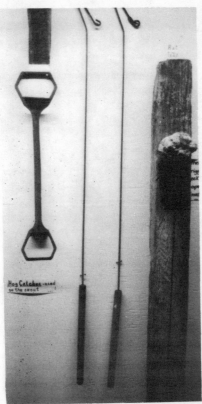

Feed grinder, hand-cranked, "Model No. 2", for corn and other grains, about 16 in. high. $16-23

Below
Feed cutter plate for display advertisement (printer's block) illustrating "Boss Feed Cutter / No. 3", 2½ x 3½ inches. $8-11

Animal catchers; metal wrench-like device center left is a hog-catcher with top handle while to right are two chicken-leg catchers. Board at far right with elevated rock is a primitive rat-trap. When bait was touched a trigger caused the rock to fall.
Photo courtesy Museum of Appalachia, Norris, Tennessee

Below
Animal yokes, including those for singles and pairs of oxen, calves and, lower left, goats. Contrary to what can be widely seen, these yokes are hung in the correct, as-used position.
Photo courtesy Museum of Appalachia, Norris, Tennessee

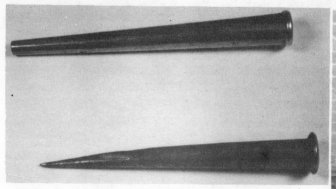

Above
"Grain thief", used to make sample grain bag contents through the fabric. Chromed steel case 4¾ in. long, hollow needle is iron with brass collar, marked "Chas. Stager & Co./ Toledo". $12-15

Farm "dolly" or grainsack truck, split handle, wooden axle and wheels, 4 ft. high. $40-50

Feed sign, paint on enamel over sheet iron, about 24 x 36 inches. Blue and yellow; rose is red. $20-25

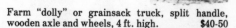

Bit box, for storing and carrying drill bits, 8 in. long; dovetailed corners. $4-7

Horse-collars, stuffed leather over forms, per each . . . $4-5

Grease gun, often used with early gasoline-powered equipment; consists of steel plunger shaft and cap with body and nozzle of four separate extruded brass pieces, 13 in. long. $17-22

Drill, hand chain; "Millers Falls", 12 in. high $33-37

Drill, mechanism; automatically feeds bit into metal $35-42

Drill, hand chain; patent information, small $22-26

Drill, chain; "Grabler & Co./Cleveland/Pat. 2-18-09" . . $35-39

Drill-bit gauge, 11 holes, wrought iron, 11 in. long $35-42.5

Drinking mug, tin, dark grey, handled $6-8

Feedbag, "Purity Feeds", white cotton, faded red lettering $4-5

Field decoy, crow, hinged wings, plywood, 14½ in. long $20-25

Firkin, (sugar bucket), large, old skim milk white paint, bentwood handle pegged in, 11½ in. high $48-55

Fire extinguisher, brass, "Fyr-fyter", with bracket, 18 in. high $29-35

Fish decoy, wood, primitive, tin fins and tail, 8 in. long . $45-60

Fishing reel, brass, with ivory handle, 3 in. long $55-65

Fishspear, four tines, wrought-iron, 13 in. long $16-20

Flax-winder, eight-armed, wood-pegged, 4 ft. high $125-150

Fleshing tool, tanner's; double-handled, 20 in. long $17-22

Flint and steel, early, hand-forged $26-30

Folk art, carved horse and rider, orig. polychrome paint, 4¾ in. high $75-95

Fork, wrought-iron, long-handled, 27½ in. long $25-35

Fork, wrought-iron, 25½ in. long $50-65

Fork, wrought-iron, two wide-set tines, 17½ in. long . . $15-20

Footscraper, cast-iron, in shape of dachshund, 25½ in. long $65-80

Footwarmer, ceramic, blue, handled, "Logan Pottery Company" $40-50

Footwarmer, punched tin, wooden frame, turned corner posts, some rust and hot-coal damage, 6x8x9 inches $50-65

Froe, blacksmith-made, orig. handle, blade 11 in. long $24-28

Funnel, filtering; copper-plated tin, "Pat'd 21 Mar 1911" $15-20

Funnel, tin, handled, 15 in. high $13-16

Funnel, handhammered copper, dovetailed, 7½ in. diameter $59.5-65

Funnel, copper, orig. tin liner, 5½x6½ inches $22.5-30

Funnel, tin, with pointed juicing spear and filter $8-10

Garden ornament, cast-iron grey rabbit, 11½ in. high .. $85-100

Gate, wrought-iron, for high fence, ornate, 34x81 inches $110-130

Ginseng digger, forged iron blade 7 in. long, iron foot crosspiece, 4 ft. straight handle................. $15-19

Glue pot, cast-iron, double-boiler, 5 in. diameter $13-15

Glue pot, copper, polished, early seaming, 3¼ in. high $30-38

Grain cleaner, "Eureka Mill/1870", incomplete, 44 in. high $40-50

Grain grinder, "#4", handle on wheel, 15 in. high $17-22

Grain stake, hand-carved, to hold top bundle in stack... $35-40

Hand-adz, metal strap-on blade, nicely carved handle, 12 in. long.................................. $60-75

Hatchet, all-brass head, break at end of one nail claw .. $22-28

Hammer, crating; "E. Bonner/1901", 10 in. long....... $13-16

Hammer, file-maker's; 9 in. long $27.5-35

Hand-vise, turn-screw closure, japanned wooden handle 7 in. long.................................. $12-16

Harness hanger, "Bubier & Co., Boston", 8 in. high $22.5-25

Harness hook, board-mounted, 3½ in. long and wide... $8-10

Harness-maker's vise, leather-covered seat, stamped "1857", refinished; small repair to jaws, 43 in. high $85-100

Hatchel, on board, heart cut-outs, dated 1853, 27 in. long $150-175

Hay knife, spade-type blade, long wooden shaft/handle. $35-40

Hay rake, wood, 68 in. wide, 70 in. long $65-75

Herb grinder, rolls between hands, round bottom dish with wear, 11 inches $45-55

Herb masher, maple, concentrics on handle, 5 in. long.. $18.5-25

Hewing dog, iron, spiked ends for holding log, 13½ in. long $11-14

Hod-carrier, "V"-shaped, on wooden pole $9-11

Hog scraper, metal ring with central wooden handle ... $4-6

Hog scraper, tin, circular blade, 6½ in. high $7-10

Hoof-leveling gauge, farrier's; "Hood & Reynolds/Boston", scale marked on brass........................ $44-50

Hook, iron, "S"-shape, wrought-iron, 5 in. long $6-9

Hook, wrought-iron, four-prong, 8 in. long........... $15-20

Horn, tin, with brass mouthpiece, 64½ in. long $105-135

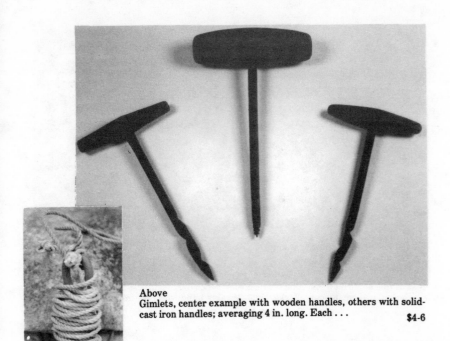

Above
Gimlets, center example with wooden handles, others with solid-cast iron handles; averaging 4 in. long. Each . . . $4-6

Left
Hook, multipurpose, hand-forged iron, with original cord. Used on farms for everything from trap drags to retrieving lost well buckets, 5 in. high. $8-13

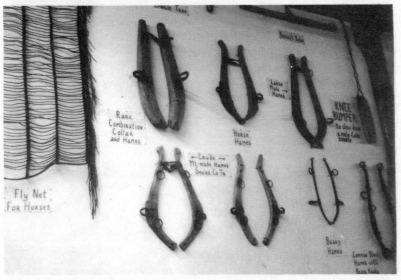

Horse harness-hames, with, at top left, a rare combination collar and hames. Far top left, a fly net worn by horses.

Photo courtesy Museum of Appalachia, Norris, Tennessee

Shaving horse (bot. left), and horseshoeing bench (bot. right) used by farrier; it retains the original equipment. Background, turning lathe with cast-iron gears.

Photo courtesy Bob Evans Farms, Rio Grande, Ohio

Harness hames, some with brass knobs at ends. Each . . . $12-15

Horse-bit, chain-type, new condition $8-10

Horse-bit, for racing, 14 in. long $10-12

Horseshoe, salesman's sample, "Snowcleat", spiked
    bottom $36-44

Horse tether-weight, cast-iron, hemispheric, 18 pounds . $24-28

Ice-cutting scene, oil on canvas, by Charles Brougham
    (1870-1951), old canvas rebacking, small repairs,
    gilded frame, 22x28½ inches $410-500

Ice pick, metal, four-pronged, 13 in. long $17.5-22

Ice tongs, iron, painted black, 15 in. long $8-10

Jar, stoneware, 3 gal. size, primitive tulip decor in
    brushed cobalt, 13½ in. high $175-225

Jar, stoneware, green running glaze, 11½ in. high $95-120

Jar, stoneware, applesauce, clear glaze on brown $29-35

Jointer, coopers; bolted to bench, double iron, "Stanley",
    35¾ in. long $85-100

Jug, pottery, Southern, unusual black glaze with blue
    highlights, 11½ in. high $22.5-30

Jug, stoneware, ovoid, 10 in. high $35-45

Jug, stoneware, cream sides and brown top, glazed,
    3-gal. size $17-22.5

Jug, stoneware, 2-gal., "A. B. Wheeler & Co./Boston,
    Mass.", cobalt bird on branch, 13½ in. high $265-290

Jug, stoneware, brown running glaze, 9½ in. high $15-20

Keg, stave-constructed, wooden bands, 8½ in. high . . . $55-70

Kentucky-style rifle, half-stock, percussion, 49½ in.
    long $300-350

Kerosene carrier, tin over glass, ventilated sides, 10 in.
    high $19-25

Kerosene can, wire handle with wood grip, 15 in. high . . $20-25

Kettle, apple butter, copper, 25 in. diameter $200-250

Kettle set, fork, brass-bowled ladle and strainer, 17 in.
    long $80-95

Kettle, salesman's sample, brass, 2¾ in. diameter $55-65

Kettle, spun brass, stamped label "H.W. Hayden's
    Patent/Ansonia Brass Co.", wrought handle,
    11½ in. diameter $55-70

Kettle stand, iron, openwork top, 13 in. high $45-55

Kneeling figure, carved wood, good primitive detail in
    hands and face, white body, modern wooden base,
    29½ in. high $475-550

Kraut chopper, wood handle, 8 in. wide $20-25

Kraut chopper, wood handle, 9¼ in. wide $25-30

Kraut-cutter, maple and ash, dovetailed hopper on
board, 10¾x36 inches $105-130

Kraut cutter, walnut, 10¼x28½ inches $75-95

Kraut cutter, hickory handle, iron blade $13-18

Ladle, carved wood, well-shaped handle, 6½ in. long . . $40-50

Ladle, hand-hammered brass, copper rivets, iron handle,
early $100-125

Ladle, hand-wrought handle, iron bowl 6 in. diameter . $37.5-43

Lamp, privy; tin, brass and tin whale oil burner, 9 in.
long $30-40

Lamp, "burglar" type; tin, clear bullseye lens, brass
crown top, orig. font and burner, 6¼ in. high $65-80

Land grant, Steubenville, Ohio, by President Madison,
1812, framed $65-80

Land grant, unstated area, by President Monroe, 1819,
framed $95-120

Lantern, candle, tin, some rust and old soldered repairs,
12½ in. high, plus ring $50-65

Lantern, tin, clear globe, orig. font without burner,
worn brown japanning, 12¾ in. high, plus ring . . . $45-60

Lantern, tin, "Empire" on base, kerosene burner altered
to accept candle, 12 in. high $30-40

Lantern, tin, whale oil burner, 5¾ in. high $130-150

Lard press, ash, with galvanized, handled pan $50-60

Latch, curved end is spring, 23½ in. long $70-85

Latch and keeper, barn-door size, 15 in. long $15-20

Leather-riveting tool, iron, lever action $8-10

Leather-worker's vise, iron sawtooth ratchet, 40 in. high . $35-45

Lightning-rod insulators, four; milk glass, two/blue, two/
white, small chips, ea. 5 in. high. The lot $55-70

Lunch bucket, tin, oval, tin food compartment, lid
drinking cup, wire bail, 9 in. long $35-45

Lunch pail, loop and bail handle, covered, 4½x6 inches $15-18

Mallet, carpenter's wood, battered burl, 15 in. long $9-15

Mallet, carpenter's, lignum vitae, hickory handle, 15 in.
long $25-35

Marionette, fully jointed wooden horse, white and black
paint, 9 in. high $40-50

Wooden wringer or press, side-crank powered, with cogged-cylinder roller against small wooden rollers. Very similar devices were once used to press wet ryestraw for basket-making. $45-55

Photo courtesy Tilson Collection

Above
Large maul, hickory handle, burl head, 30 in. handle.

$12-16

Very early rural mailbox, hand-shaped metal, small door with upper hinges.

Photo courtesy Museum of Appalachia, Norris, Tennessee

Heavy-duty plow with two large plowshares throwing earth out from center. Maker-symbol that appears to be a joined "IH" inside a large "C". Plow may have been for ditching, firebreak or agriculture.

Photo courtesy of Hubbell Trading Post National Historic Site, Ganado, Arizona

Walking plow, turn-of-century Kenwood type, maker-marked as follows: On mold-board, "AVERY 14-M-L"; on beam, "P-3 ILL. G.".

Photo courtesy of Hubbell Trading Post National Historic Site, Ganado, Arizona

Above
Paint scrapers; top, factory-made device with pistol handle, marked "Hook Scraper No. 75", 7 in. long.
$11-14

Bot. left, hand-forged iron scraper, just under 8 in. long. $3-4

Fine farm press, 16 in. wide. Oak hand-cut base, forged iron frame, wheel and pressing face, chamber of cast and milled brass. From Pennsylvania. $125-150

Left
Farm press, tripod-legged, tin inner chamber, wrought screw center column, two-handled turn.
$22-27

Hog-ring fastener or pliers, head with three slots for two sizes or double-ringing. "Pat'd Sep 8 (18)74", 7 in. long.
$7-10

Pioneer home with several accoutrements. Left, large iron cooking kettle suspended over fire. Center, lower, triangular coop for hen and chicks.

Photo courtesy Museum of Appalachia, Norris, Tennessee

Walking plow, cast plowshare, wooden tongue. $40-50

Marionette, fully jointed wooden horse, primitively
    carved, leather saddle, tape halter, cloth and carpet
    mane and tail, 20 in. long, 24 in. high $500-600

Masonary ties, pair, cast-iron oblong foliage plates,
    peeling paint layers, 9½x15¼ inches; the pair $45-65

Maul, wood, heavy head, short handle $10-13

Measure, bellied, pewter, marked "pint" $165-190

Measure, bentwood, old green paint, 5¼ in. diameter . . $20-25

Measure, pewter, with brass lip, 2½ in. high $45-55

Measuring tool, cooper's, iron with wood handle, 20½ in.
long $40-50

Meat hook, wrought iron, four prongs, 6 in. high $55-65

Meat press, "Wilder's", tin, wood pressure plate $35-40

Meat rack, curved wood, raised in center, designed to
fasten against wall. With 16 hooks, 8 ft. long $250-275

Micrometer, 1 to 2 inches, "J.T. Slocumb/Prov. R.I.",
5½ in. long $14-18

Milk can, 15 gal., copper name-plate $19-24

Milking stool, four-legged, central hand-grip hole, 14 in.
high $22-26

Mill pick, used for dressing and "freshening" grooves in
millstones, iron head, 9 in. long $24-29

Miniature scene, barn with wood stalls and block animals,
whole 12 in. high $30-40

Miter box, adjustable, on legs $45-50

Mortar, brass with crack and pestle, in primitive wooden
hanging frame, 14½ in. high $65-75

Nail claw, metal, 14 in. long $9-13

Nail holder/guide, "Never Pound Your Thumb Again" . . $9-12

Noise-maker, cogged wheel against tensioned wooden
leaves, 8½ in. long $12.5-16

Noise-maker, 16 in. long, two hardwood leaves $24-29

Nutcracker, wood, threaded screw, 5 in. long $12.5-15

Odd jobs tool, "Stanley #1/Pat. 1-25-87", 4½ in. high . . $28-45

Oil can, brass, for steam engine cylinder oil, 9 in. long . . $24-30

Onion rake, wrought-iron fittings, beautifully done,
32 in. wide, handle 69 in. long $120-145

Ornament, for top of flag pole, 9¾ in. high $25-30

Pail, spun brass, wrought-iron handle, "Hayden Patent",
bottom with small holes, 12 in. diameter $47.5-60

Percussion caps, tin box of 100, copper lid, ca. 1873 . . . . $8-12

Plane, plow; beech and boxwood, brass trim, "Porter A.
Gladwin/Boston, Mass.", ca. 1860, brass screwstop
lever $135-150

Plane, rabbet; wood, "Philad'a/Warranted/D.H.", re-
placed wedge, 11 in. long $30-35

Plane, rosewood face, "Stanley #81" $18-23

Pot, cast-iron, for lead-melting, wire-bale handle, 6 in. diameter $14-18

Potato carrier, pine, trough with end handles, 10½x21x78 inches $25-35

Powderhorn, small, primitive engravings of birds, flowers, etc.; "G.R. Grant", 5⅝ in. long $80-95

Press, all-wood, crank-type with wood screw, 42 in. high $255-280

Pruning saw, curved duo-edged blade, two tooth sizes, wood handle 16 in. long. $13-16

Quail decoy, wooden, tin whistle inside with connected tube, old worn paint, 8½ in. high, unusual $350-400

Rabbit, carved wood, weathered, age cracks, 15 in. long $125-150

Rake, all wooden, pioneer, 69 in. long $60-70

Reamer, wheelwright's; to enlarge hub holes, hand-forged, 16 in. long $32-37

Reaping hook, "T.W. Shaws Co./Globe Works", 19 in. long $30-35

Reloading tool, "Winchester cal. 40-60 W.C.F/Pat. Sept. 14, 1880". $23-27

Ring, with four double hooks, 6 in. diameter $50-60

Rope-making outfit, metal, handle-turned holed disc . . . $20-25

Rope-twisting machine, board-mounted $35-45

Router, coachmaker's; maple or beech with 9½ in. brass sole plate and brass fittings on screw control and router top $115-130

Rule, shrinkage; "Kuffel & Esser Co./N.Y.", brass ends, 24 in. long. $23-28

Sap bucket, four-finger lap, twist hoops $28-33

Sap yoke, (maiden yoke), one-piece wood, for maple-sap buckets $40-45

Sausage filler, "Sargent & Co./N.Y.", iron and tin. $30-36

Sausage grinder, wood, crank handle, box size 6x6x10 inches $85-100

Sausage stuffer, wood, base slab, wooden lever arm 30½ in. long, 13 in. high, with tin plunger and wooden mallet $82-95

Saw, keyhole; handle repair, old, 9½ in. long $10-13

Saw, mitre; early, 7x35 inches $35-42

Saw, pruning; cast-iron frame $24-30

Saw, frame; rectangular wood frame 30 in. long $28-32

Saw, back; wooden handle, old, 17 in. long $7-9

Saw, vet's; dehorning, metal, 16 in. long $10-13

Saw-set, metal, correcting tooth angle, pliers grip, screw-adjustable head $8-11

Scale, cast-iron, tin-plated brass pan, "PerfectionScales/American Mfg. Co./Philadelphia", 15 in. long $47.5-60

Scoop, food or grain, round wooden handle, 13 in. long $14-18

Screwdriver, brass ferrule, "Peck & Hickman, Ltd./London", 9 in. long $15-19

Scribe, marking; wooden, adjustable gauge and handle, 20 in. long $21-28

Scrub board, wood, sawtooth surface, 3½x27¼ inches . $29-35

Scorp, cooper's; blade on forked wood, 17 in. long...... $38-45

Scorp, woodworking; "Norway/1865", 6 in. long $28-33

Scythe, handwrought, rootwood handle, 38 in. blade .. $17.5-25

Seeder, "Tiron", handcranked at waist level to space seeds $23-28

Sled, wooden including runners, paint traces, 30 in. long $45-55

Sled, about 19x35 inches, all-wood, 6 in. wooden railing. New England, ca. 1870, used to pull food from winter storage to house kitchen, unusual $85-100

Shoe last, cast-iron, black paint, 8¼ in. high $25-35

Shoemaker's tools; two, incomplete wooden boot last 19 in. high/iron tool (marked "Lightning/Fulton, Ill.") for stretching shoes. The two $35-45

Shoemaker's tray, cast-iron, revolving, "Star Nail Cup", 9½ in. diameter $20-25

Sickle, replaced socket handle, old $11-14

Sickle, wood handle, iron blade, 21 in. long........... $12-18

Sickle, wrought-iron, stamped "I. Christ", 22¾ in. long $32.5-40

Sifter, wire screen, bentwood rim, 17½ in. diameter ... $20-26

Singletree, wood and iron, fastened behind horse, 14½ in. long $12-15

Skimmerhorn, flattened and pressed, small holes in bottom, 7 in. long, old $16-20

Skimmer, brass, delicate punched design in bowl, wrought handle $100-125

Saw collection. Left, crosscut saws; top left and center, hand saws; top right, frame or felloe saw; bottom center, circular saw; bottom left, bow saws.

Photo courtesy Museum of Appalachia, Norris, Tennessee

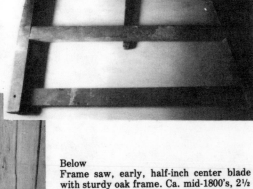

Veneer saw, with twisted-cord blade-tightening arrangement; small size, about 13 in. high.
$16-20

Above
Sled, about 23 in. high. This is a scarce late-1800's variety (in old red paint) known as a chair-sled or seat-ski. Bottom runner is 31 in. long. $85-100

Below
Frame saw, early, half-inch center blade with sturdy oak frame. Ca. mid-1800's, 2½ ft. high. $24-30

269

Long barnyard sawhorse, three "X"-braces, with log in position. Background, stacked "splits" of wooden shingles.

Photo courtesy Museum of Appalachia, Norris, Tennessee

Sled, iron-clipped runners, braced-wire frame, wood top, ca. late 1800's.

$120-150

Courtesy private collection

Early manure spreader, rear view showing spiked throwers mounted in wooden beams; note all-metal wheels.

Photo courtesy of Hubbell Trading Post National Historic Site, Ganado, Arizona

270

Spigots, for barrels or kegs, all wood.　　Each . . . $5-8
Photo courtesy Fairfield Antiques, Lancaster, Ohio

Skimmer, brass, riveted with copper to iron handle, 16 in.
　long $95-115

Skimmer, copper, punched bowl, wrought-iron handle . $80-100

Skimmer, wooden, 4 in. bowl with 3½ in. handle $29.5-35

Slaw cutter, walnut, heart cut-out, 7¼x18 inches $27.5-35

Sled, wood, metal-tipped runners, worn red paint, old
　repair, 37 in. long $55-65

Sled, wood, iron braces and wrought-iron runners, 52 in.
　long $50-60

Sleigh, wooden, metal-tipped runners; worn green paint
　with transfer of running horse, 10½x30½ inches . . $105-125

Sleigh bells, five, on leather strap 15½ in. long $50-60

Sleigh bells, 24 graduated-size bells, strapped $65-80

Slick, (bark spud), steel, 3½ in. wide, "T.H. Witherby",
　31 in. long $65-75

Snow knocker-hammer, to tap snow from horses' hooves . $25-30

Snowshoe thong cutter, Maine, handmade and pegged,
　forged table clamps, 26 cutting blades, two 15 in.
　handles, 28 in. long $70-85

Spokeshave, handmade, old, 11 in. wide $18-22

Spokeshave, double; "Seymour Smith & Son", 9½ in.
　wide. $19-24

Spring latch, 7½ in. long $30-35

Stake, tinner's; blow-horn, metal, 24 in. long $55-65

Stitching head, saddle-making, 15 in. long $16-20

Stove, box; four-footed, cast iron, ash platform, overall
    size, 13x18x25½ inches $150-175

Stove, "Midget", removable top for boilers, "Atlanta
    Stove Works" $145-170

Strawberry huller, "Nip-It", metal $3-4

Strawberry carrier, wood platform for six 1-qt. boxes, top
    center carrier handle $22-28

Strike-a-light, wrought-iron, tooled edges, stamped signa-
    ture "Ollison", 5¼ in. long $102.5-130

Square, angle; "B. Stone", folded, 14 inches $12-15

Square, angle; "Star Tool Co./Pat. 1867", folded,
    8 inches $16-19

Sugar bucket, stave-constructed, wire bale, wood handle,
    8¼x8 in. high $75-90

Sugar bucket, stave-constructed, wooden, 12½x12 in.
    high $55-70

Sump mortar, hollowed from one log, maple, 21 in. high $175-200

Surveyor's reel, early, 7¾x18 inches $14-18

Swift, whalebone and ivory, red and black scribe line,
    15 in. long, for yarn-winding $495-550

Tap, for cutting wood threads 2¼ in. across; 12 in. long $65-75

Tar bucket, Conestoga wagon, wood, leather thongs,
    9½ in. high $35-45

Taster, wrought-iron and brass, 11¾ in. long $75-90

Taster, wrought-iron and brass, 9 in. long $125-150

Thumb latch, with latch rod and staple, good detail, 13½
    in. long $75-85

Thumb latch, double heart finial, thumb piece missing,
    11¼ in. long $35-45

Thumb latch, tulip on each end, thumb piece alongside,
    16 in. long $135-150

Tin, "Mica Axle Grease", one-pound size $4.25-6

Tin, "AA Saddle Soap", 3 in. diameter $5-7

Tinner's bench, with some attached tools $110-125

Tinker's tool box, orig. green paint, rings for shoulder
    straps $47.5-55

Tobacco cutter, cast-iron, "Red Tin Tag", 17½ in. long . $35-45

Left
Farm tools; top, pruning saw, 15 in. long, hardwood handle. $6-8

Bottom, mattock, part of original label, fine condition. $8-11

Above
Tools, for working stone or brick, example to left with wrought-iron head, 12 in. long.
                                                                    $8-11

Right, cast-steel head with pointed tip and right-angle blade. $9-12

Right
Tool, known sometimes as a crate-opener or wrecker's bar. Cast-iron metal, hardwood side-handle, plus hammer, hatchet and nail-pry; about 14 in. long.
                                                $13-16

Triple-tree, used to hitch three horses to equipment.

Photo courtesy Bob Evans Farms, Rio Grande, Ohio

273

Very large meat-salting trough, hand-scooped from tree trunk.

Photo courtesy Museum of Appalachia, Norris, Tennessee

Blacksmithing tools.

Top, hammer 13 in. long, head with cutting edge and pointed tip $7-10

Center, head only, hand-forged, well-balanced. $3-5

Lower, hammer, cutting edge and pounding end. $8-11

Farrier's tools, bottom specimen 7 in. long.

Bottom, hoff-knife, bone handle $7-10

Top, farrier's horseshoeing hammer. $9-12

Large pioneer weaving loom, of the type used in the loom-room or loom-house.
Photo courtesy Museum of Appalachia, Norris, Tennessee

Tobacco cutter, lever-action, "T. C. Johnson/Quincy, Ill." $19-24

Tobacco leaft cutter, blade, shaft and wood handle, 10 in. long $12.5-15

Tool box, oak, tray with five compartments, 6¼x9½x18 inches $60-75

Tool box, script "Tools", lidded, 4x5x12 in. long $22-26

Tractor box, raised lettering, "Richard Mfg. Co." $19-25

Trade card, "E. D. Murphy/Harness Maker", from Concord, New Hampshire...................... $2-3

Transit level, brass, "W. & L.F. Gurley/Troy, N.Y.", 13 in. long $90-120

Trap, single-spring, orig. pan, chain and hooks, 5 in. jaws $30-35

Turnscrew, "Sheffield and Broad Arrow Combination Plier & Cutter" $24-29

Try square, copper, rare, 12 in. long................. $22-28

Try square, brass bound, iron measure 6 in. long....... $12-15

Utensils; three, wrought-iron, two skimmers and fork, 15, 17 and 18 in. long. The lot.................... $35-45

Vegetable slicer-shredder, wooden, clamps to table, hand-crank on top, 19 in. high $35-45

Vise, bench-type, 9 in. high $8-10

Vise, metal, handheld, 4½ in. long $9-11

Vise, saw; clamp type, 8 in. high $9-12

Vise table, harness-maker's, wooden pressure handle, 28½ in. high, ca. 1860 .... $30-35

Wagon seat, pine, worn old red paint, 11x28¾x15½ in. high $250-300

Wagon-tire bender, wood, unusual, 38 in. long $65-80

Wagon-wheel brake-skid, shoe-shape, cast-iron $27.5-32

Water-color on paper, primitive Pennsylvania farm scene, by "Hattie Brunner, '63", framed, 14x18 inches ... $1550-1880

Water cooler, stoneware, impressed label "The Kenton Cooler", corked spigot hole, rim hairline, 12½ in. high $40-50

Water cooler, stoneware, eight-gal., double handles with two blue brushmarks, and "8"; 19¼ in. high $55-70

Water jug, pewter, screw cap on spout, "Flagg & Homan", 10¼ in. high $35-45

Weathervane arrow, cast-iron, ruby-flashed glass inset with moon and comet, new finial and stem, 22½ in. long $65-80

Weathervane figure, copper bull, cast zinc head, wrought-brass horns, dark patina; some repairs, bullet holes, 30 in. long $1800-2100

Weathervane figure, early, wood full-figure rooster, old damage to comb and tail, weathered red and blue paint, 19½ in. high $2850-3300

Weathervane, sheet metal, rooster; old worn red, white and gold paint, 29 in. high $145-170

Wedge, powder and cap, for log-splitting $46-55

Wedge, splitting, marked "WPA", 14 in. long $21-26

Wheelbarrow, all wood, red paint, orig. stenciling, wood-spoked wheel $200-235

Windmill weight, cast-iron, 17¼ in. long............. $200-245

Wire-winder, wooden spool, crank handle $42-48

Wooden boot forms, the pair $17.5-22

Wooden horse, primitively carved, old paint, 3½ in. high $20-24

Wool carder, primitive, wrought-iron spikes, wooden handle, 12 in. long $35-42

Wrench, buggy; brass, double-headed, 8½ in. long $18-23

Early water system; roof in background drains into well; elongated well bucket is raised and lowered with crank and windlass arrangement.

Photo courtesy Museum of Appalachia, Norris, Tennessee

Wheelwright's area, with a dozen examples of wheel hubs; good hubs required fine woodworking and blacksmith skill to be well "ironed".

Photo courtesy Museum of Appalachia, Norris, Tennessee

Wire head of coals shovel, wooden handle 3 ft. long.
$23-28

Wrought-iron ladle, made from iron pipe giving a hollow handle and loop. Piece is 2½ ft. long, with duo-spouted dipper, for use with buterching kettles, maple-syrup making, and so forth. $12-15

Collection of farm-related items, all from Maine.

Left, large axe with head about 1 ft. high, old handle. $90-120

Center-right, plow-plane, ornate "iron"-holding wedge. $42-50

Top right, set of four marking scribes, wood turn-screws. Per each . . . $12-18

Collecting Farm Items—Some Considerations

These words are not directed toward the advanced farm-item collector, for that person in all probability has already established specific sources for whatever is sought. Instead, this is a brief summary of where the first-time collector might begin, and offers a look at some options the intermediate collector might have passed by.

While there are indeed hundreds of classes of farm items, some places seem to be better than others for obtaining good pieces at very reasonable prices. And—apologies in advance to the hard-working owners and managers—big-city antiques shows, sales and shops do not seem to be where much time should be spent. Perhaps it is obvious, that for farm items, go to farm areas.

In general, city shops and the well-publicized shows and sales tend to concentrate more on traditional collecting fields, rather than what is upcoming. Farm items are in this category, with collector interest only now picking up momentum. Selections are not always large, and prices tend to be toward the higher end of any value range. Quality is usually high, but true availability of such pieces to the collector with limited means is questionable.

Instead, it is suggested that the collector become familiar with four other potential sources. They are: Selected flea markets, out-of-the-way antiques shops, country and small-town auctions, and yard sales. This is partly a push to go where many others do not, but more a hint to go where the farm antiques and collectibles are.

Flea markets—those large outdoor and/or indoor associations of dealers and sellers—have, frankly, various levels of social intercourse and quality of offerings. Decide where you feel comfortable, what you can afford, and set up a calendar-plan that lets you visit those you consider important. The term "flea market" itself is unfortunate, an out-moded description from Europe, with some negative connotations. Perhaps such gatherings might better be termed "collectors' fair", or something similar.

At any rate, farm items generally abound in "as-is" condition, and it's a real ground floor for those who seek even scarce or unusual pieces. While some items are under-valued, others over-valued, the vast middle pricing structure makes visiting worth-while. The sheer size of many such markets means that thousands of items can be viewed and compared. Bargaining and trading are not unheard-of. Some dealers do not wish to accept checks, even with the best identification, but cash is always welcome.

Out-of-the-way antiques shops in rural areas tend to be real treasure-houses. With low upkeep and large display spaces, they can cram in ten times the volume of a city shop, and keep prices at welcome levels. In New England, for example, entire barns have been turned into such shops. Often materials from a number of nearby farmsteads, via estate sales, can be sorted through.

The inventory of such shops seems to revolve quickly, with much new input and material every month or so. The beginning collector would do well to become known to the owner, and state whatever needs clearly.

Country and small-town auctions have perhaps been over-rated as bargain-sources, and a newcomer might be surprised by the prices paid for certain antiques. Mostly, such bidding is done by two widely divergent classes, those who really know what they are doing, and by inspired bidders caught up in auction fever.

One very educational aspect of such auctions is seeing how much dealers are willing to pay for certain farm items. Overall, the competition—hence, the upward bidding and the final price—is often less at some of these smaller auctions.

Many Sunday papers (classified section) and trade periodicals carry such auction listings, and one could do worse than attend those with only a few farm items listed. Sometimes the number of items to be sold is several times as great as those actually advertised.

Yard and garage sales have drawbacks, in that one must use expensive gas and considerable time to take in half a dozen of them. Most are not pure-antiques offerings, but accumulations to be converted to cash.

But, amid all the clutter, good farm pieces regularly do turn up. Usually, no one knows where they came from, and prices may be somewhat curious, but a determined collector can obtain some surprising and gratifying finds for a few cents on the dollar.

This sums it up: Good farm items and below-market prices do exist. But they rarely come to the collector. The collector must seek them out.